Cryptic Concrete

# RGS-IBG Book Series

For further information about the series and a full list of published and forthcoming titles please visit www.rgsbookseries.com

## Published

# Cryptic Concrete

*A Subterranean Journey Into Cold War Germany*

Ian Klinke

**WILEY** Blackwell

*Registered Office(s)*
John Wiley & Sons, Inc., 111 River Street, Hoboken, NJ 07030, USA
John Wiley & Sons Ltd, The Atrium, Southern Gate, Chichester, West Sussex, PO19 8SQ, UK

*Editorial Office*
9600 Garsington Road, Oxford, OX4 2DQ, UK

For details of our global editorial offices, customer services, and more information about Wiley products visit us at www.wiley.com.

Wiley also publishes its books in a variety of electronic formats and by print-on-demand. Some content that appears in standard print versions of this book may not be available in other formats.

*Library of Congress Cataloging-in-Publication Data*

Names: Klinke, Ian, author.
Title: Cryptic concrete : a subterranean journey into Cold War Germany / by Ian Klinke.
Other titles: Subterranean journey into Cold War Germany
Description: Hoboken, NJ : John Wiley & Sons Ltd., [2018] | Series: RGS-IBG book series | Includes bibliographical references and index.
Identifiers: LCCN 2017051957 (print) | LCCN 2018004340 (ebook) | ISBN 9781119261131 (pdf) | ISBN 9781119261124 (epub) | ISBN 9781119261032 (cloth) | ISBN 9781119261117 (pbk.).
Subjects: LCSH: Nuclear weapons–Government policy–Germany (West) | Geopolitics–Germany–History–20th century. | Biopolitics–Germany–History–20th century. | Bunkers (Fortification)–Germany (West) | Guided missile bases–Germany (West) | Civil defense–Germany (West–History. | Military maneuvers–Germany (West) | Nuclear warfare–Government policy–Germany (West) | Landscapes–Germany (West) | Cold War.
Classification: LCC U264.5.G3 (ebook) | LCC U264.5.G3 K55 2018 (print) | DDC 355.02/17094309045–dc23
LC record available at https://lccn.loc.gov/2017051957

Cover Design: Wiley
Cover Image: Image of NATO nuclear weapons storage site, West Germany © Ian Klinke

Set in 10/12pt Plantin by SPi Global, Pondicherry, India

The information, practices and views in this book are those of the author(s) and do not necessarily reflect the opinion of the Royal Geographical Society (with IBG).

Printed in Singapore by C.O.S. Printers Pte Ltd

10 9 8 7 6 5 4 3 2 1

# Contents

# Series Editor's Preface

The RGS-IBG Book Series only publishes work of the highest international standing. Its emphasis is on distinctive new developments in human and physical geography, although it is also open to contributions from cognate disciplines whose interests overlap with those of geographers. The Series places strong emphasis on theoretically informed and empirically strong texts. Reflecting the vibrant and diverse theoretical and empirical agendas that characterise the contemporary discipline, contributions are expected to inform, challenge and stimulate the reader. Overall, the RGS-IBG Book Series seeks to promote scholarly publications that leave an intellectual mark and change the way readers think about particular issues, methods or theories.

For details on how to submit a proposal please visit:
www.rgsbookseries.com

David Featherstone
*University of Glasgow, UK*
**RGS-IBG Book Series Editor**

# Preface

It must have been in 1990 when I found out that the hill behind my friend's sandpit was hollow, perhaps so hollow that it could swallow an entire army. What lay beneath the vineyards was a secret, but it was one that was passed on from child to child. The story that circulated amongst us was one of a subterranean city with streets and lanterns, buses and cars, bakeries and sweet shops, as well as tanks, missiles and soldiers. My friend and I gazed with enthralment at the barbed wire, the guards and the watchtowers, behind which we correctly suspected lay the entrance to this secret underworld. It made me feel uncomfortable – and yet my imagination was drawn to it. As we dug holes deep into the sandpit for our plastic soldiers, missiles and tanks, we lost ourselves in geopolitical fantasy. We were child strategists, subterranean generals, standing tall at the end of history.

Of course, the meaning of the events that brought an end to the Cold War had not been lost on us. The political excitement was palpable, the feeling of cultural superiority overpowering – even for an eight-year-old. Soon, our television set would show a city plunged into an extravagant display of green fireworks. This city was Bagdad and the green light was the flickering of the Iraqi anti-aircraft guns in their attempt to resist the world's sole remaining superpower. It was a truly captivating display of power – though it was ultimately as intangible as the nuclear explosion I had once seen in an American film. I found it difficult to relate to on a personal level. The West German government's nuclear bunker behind my friend's house, however, was something much more concrete and tangible. The bunker felt so real, even though it was so well concealed. It seemed to be mine, even though it was never made for mere mortals like me. For me, this concrete survival shell was a forbidden land of plenty and a place of salvation, a place where my fantasies were safe. This secret and sacred space was uncanny – *unheimlich* – in the Freudian sense of something that is both alien and familiar, repulsive and attractive.

The social theorist Paul Virilio has argued that bunkers have functioned as underground places of security, hidden and forbidden places, 'as in the English word, cryptic' (Virilio, in Armitage 2009: 23). Reflecting on German air-shelters that were converted into churches, he has suggested that 'these places of shelter from danger, and places of worship, [] are also places of salvation' (ibid.). As in the stone chambers beneath Christian churches, death has a haunting presence in the nuclear bunker. The bunker is an ambivalent space, both 'shelter' and 'grave' (Bennett 2011a: 156), 'womb' and 'tomb' (Beck 2011: 82). Part of this ambivalence may be inherent in the very material of which most bunkers are made. As Adrian Forty (2012: 169) puts it:

> Concrete is a base material. Its dense mass lends it to the resistance of forces, whether natural or man-made. Good for foundations, sea defences, fortifications, nuclear shields, anywhere that monolithic inertness is called for, this quality puts it low down in the hierarchy of materials. At the same time, though, concrete has from its earliest days appealed to church builders.

Forty reminds us that despite its ability to withstand tremendous forces, concrete holds an ambiguous position between the modern and traditional, cultural and natural. Notwithstanding its success in the twentieth century, concrete is neither a modern creation (for its use by humans dates back thousands of years) nor indeed a purely cultural phenomenon (it does exist in natural form despite being rare as such). Concrete, in other words, is difficult to categorise – it almost wants to be interpreted.

Twenty-five years later, I understand why the bunker and its surroundings had exerted such a strong attraction on me, for the infantile war game that I had once played behind my friend's house had a very particular personal significance. From an early age I had been forbidden to play with toy soldiers and other such 'symbols of militarism'. Mine was a life without the symbols of boyish masculinity. In my school friend's house, right next to the government's nuclear bunker, however, the rules were different and I was able to indulge in war games of all sorts, from the positioning of plastic tanks in a sandpit to more elaborate strategy games. I knew I had to remain silent about this forbidden form of enjoyment when I returned home – but this did not in any way compromise my guilty pleasure. Only decades later, when I started to develop an academic interest in the site that lay behind the sandpit, did I find out that there had been similar subterranean and forbidden games going on underneath our playground. For every two years, the West German state would lock up its political and military elites underground so that they could play the apocalypse. I now understand that these games were in fact driven by an obsessive politics of earth and life.

Geopolitics, the politics of earth, was first developed at the turn of the twentieth century as a geographical theory about state behaviour. It posited that states needed to conquer and dominate political space if they wanted to survive in a competitive international environment. In Germany, this geographical discourse was crucial in legitimating the Third Reich's invasion of the Soviet

Union and the conquest of living space in Eastern Europe. Geopolitics was always intertwined with biopolitics (the politics of life), the belief that the state should be understood as an organism that struggles for survival. The Third Reich's extermination of unwanted populations became possible only by branding some groups as cancerous cells within this organism. These two forms of power were linked in many ways, but most crucially in the fantasy of conquering *Lebensraum* (living space) in Eastern Europe. This fantasy, or so I argue in this book, did not simply vanish with the demise of the Third Reich but took on a new form. In order to understand this fully, we need to grapple not just with the strategic discourses of the Cold War but with the violent architecture of the atomic age itself.

This book then prompts us to reassess the history of geo- and biopolitics by exposing the ways in which the Cold War reproduced and inverted the spaces of survival and extermination that had emerged in and through World War II. In an attempt to understand the material architecture that was designed to protect and take life in nuclear war, I explore two types of structure that stood at the vanishing point of geo- and biopolitics – the nuclear bunker and the atomic missile site. Analysing a broad range of archival sources through the lens of critical theory, I argue for an appreciation of the two subterranean structures' complementary nature. Following Eyal Weizman, I approach architecture as solidified political forces, or 'politics in matter', matter that we can study through its form and ornamentation, as well as the organisation and infrastructure that enables and sustains it (Weizman 2007: 5–7). The architects of these violent geographies are thus the military strategists, engineers, civil defence planners and politicians of the Cold War state. But unlike Weizman (2002: 2), who sees geopolitics as a 'flat discourse' which fails to comprehend the three-dimensionality of modern warfare, I argue that these male strategists, and they are almost exclusively men, were already animated by the idea that Cold War geopolitics had to be fought in three-dimensional space, specifically, of course, in subterranea.

By examining the politics of nuclear weapons in West Germany in both an intellectual and an architectural register, *Cryptic Concrete* thus makes a tentative step in the direction of a biopolitics of the Cold War, an issue that was recently proposed by Collier and Lakoff (2015; see also Klinke 2015). The book also seeks to contribute to recent and ongoing debates on the materiality of geopolitics by interweaving the analysis of material forms with an examination of geo- and biopolitical thought, revealing how military architecture remained in dialogue with these ideas even after they had been proclaimed dead.

The Federal Republic of Germany is an excellent starting point for any such investigation because of the country's role as a designated battlefield in the case of a war with the Warsaw Pact. After joining NATO in 1955, Bonn participated in and drove the alliance towards a policy of hard-line nuclear deterrence. This policy and the subsequent nuclearisation of West German territory meant that the country permanently played with the idea of national suicide in ways that invoked in unambiguous terms the final days of the Third Reich. Indeed, the West German examples can be used to illustrate some of the historical continuities between the fascist and the Cold War state, not least because there was a vast overlap of personnel, ideology and military technology before and after 1945. By looking at Germany, the book attempts to turn academic debates on military landscapes away from their Anglo-American bias to reveal the shared origins of fascist and Cold War geopolitics. Through tracing the emergence of the Cold War, I hope we can learn to appreciate that 'the detonation of the first atomic bomb' did perhaps not mark 'the end of one kind of time, and the apotheosis of another' (Masco 2006: 1). Rather than 'explod[ing] experiences of time [and] undermining the logics of the nation-state' (ibid.: 12), the West German Cold War state found new ways of articulating a very familiar biopolitical modernity in which the state fostered some forms of life and abandoned others, valorising some deaths and failing to remember others. In West Germany, the technological move from Dresden to Hiroshima, or what Peter Sloterdijk (2009: 57) calls the shift from thermoterrorism to radioterrorism, was in fact framed through and organised around similar leitmotifs as Nazi geo- and biopolitics, including the latter's obsession with questions of survival and extermination.

Despite its ambition of contributing to theorisations of both geo- and biopolitics, this is also of course a book about Germany. *Cryptic Concrete* is intended for readers who want to understand the history and politics of the West German state, which remains of course the legal basis for a reunified Germany. This book is not, however, in any way meant to form a comprehensive study of German Cold War history, nor does it claim to have unearthed any particular sites that were previously unknown to the public. Instead, it sets out to re-read Germany's nuclear landscapes through critical theories of geo- and biopolitics. In doing so, the book is foremost a geographical one that tries to understand how ideas about space, power and survival, developed by the likes of Friedrich Ratzel and Karl Haushofer in the late nineteenth and first part of the twentieth century, managed to survive the demise of National Socialism. It tells the story of how the proto-fascist idea of the state as an organism produced particular architectural forms, not just in the Third Reich but also during the Cold War.

The starting point and underlying assumption of this book is that power operates in material as much as ideational ways. Rather than abandoning the study of geopolitical traditions alongside the overly textual focus that marked political geography during the 1990s, *Cryptic Concrete* tries to develop more imaginative ways of thinking with and against the geopolitical tradition. In doing so, it starts from the premise that whilst the study of geopolitical texts has a tendency to be merely 'parasitic' upon a particular form of writing (Ó Tuathail 1996: 53), more recent scholarship, which has tried to rethink geopolitics along 'more-than-human' lines, often loses sight of its object of study. Whilst the former runs into the danger of being only concerned with the 'mummified' remains of what was once an influential mode of thought (Ó Tuathail & Dalby 1998: 2), the latter either treats geopolitics as a mere synonym for global politics or develops a conception of geopolitics that is strangely detached from any previous understandings of the term. Instead, this book tries to fuse the study of geopolitical traditions with the study of the ways in which geopolitics is forged through the built environment and imprinted onto the human body. It is thus interested in the places in which geopolitical subjects are formed.

In doing so, a few words of caution are imperative. Whilst the book does feature a detailed discussion of geopolitical thinkers, I neither wish to reduce the history of geopolitics to the ideas of important men (for this critique, see Sharp 2000a: 363) nor do I intend to overstate their direct influence on political events. Rather, I would like to argue that we can find an obsession with spaces of national survival and extinction – which was first powerfully articulated through German geopolitics in the late nineteenth and early twentieth century – in the military landscapes of the Cold War. Geo- and biopolitics, in other words, managed to survive despite having been proclaimed dead, a resilience I have described elsewhere as 'undead' (Klinke 2011: 719; see also MacDonald in Jones & Sage 2010). In this way, then, the book is meant as a provocation to reconsider the politics of nuclear war through notions of geo- and biopolitics. This, I feel, is necessary because the Cold War is often regarded today as a historical absurdity with little relevance for the contemporary world. It is especially phenomena such as bunker tourism that often deny 'the origins of these sites in periods of deep ideological violence, ruination and trauma' (Graham 2016: 359). In her history of the US intercontinental missile 'Minuteman', Gretchen Heefner remarks:

> We are told that nuclear deterrence is no longer our bulwark against global conflict. We have stepped back from the precipice. Our children no longer wake from nightmares about thermonuclear war. Our contemporary nuclear fear, a dirty bomb or rogue nuke, is of a different variety, less cataclysmic. The massive armaments of deterrence do not make sense in the age of global terror. Not even the people of South Dakota who lived for so long with the Minutemen seem to want to tell the missiles' stories. But these silences and spaces are as misleading as the quiet that fell over the missile fields during the Cold War' (Heefner 2012: 4).

Heefner continues by noting that around half of the Minuteman missiles that were deployed in the 1960s remain in service today, 'still capable of reaching targets around the world as quickly as you could have a pizza delivered to your door' (ibid.: 5). But it is not simply necessary to be aware of the continued presence of nuclear weapons. It is also important to understand them as embedded within geopolitical logics that to some degree *predate* these weapons. If we want to make sense of the missile silo and its counterpart, the bunker, then we need to take seriously the history of German geopolitics, a story that did not, as sometimes assumed, simply end in 1945, for the fantasy of conquering *Lebensraum* managed to survive precisely by going underground.

This book is informed by over nine years of research on German geopolitics and the Cold War. The empirical research was conducted during my postdoctoral fellowship at the University of Oxford (2013–2015) and I should therefore like to thank the School of Geography and the Environment for giving me the opportunity to pursue this project. The book was finished after I had taken up an Associate Professorship at the University of Oxford, in association with St John's College.

It is difficult to acknowledge everyone who has had an input into this book. Key interlocutors have included the like-minded bunker enthusiasts Brad Garrett (University of Sydney), Luke Bennett (Sheffield Hallam) and Silvia Berger-Ziauddin (Universität Zürich), fellow researchers of geopolitics such as Mark Bassin (Södertörn Högskola) and Gonzalo Pozo (King's College London), the architectural historian David Haney (University of Kent) as well as my old friends Ludek Stavinoha (University of East Anglia) and Caspar Richter. I would like to thank in particular my colleagues Gruia Badescu, Maan Barua (now University of Cambridge), Colin Clarke, Patricia Daley, Joe Gerlach (now University of Bristol), Britain Hopkins, Derek McCormack, Fiona McConnell, Tim Hodgetts, Craig Jeffrey (now at the Australia-India Institute), Thomas Jellis, Kärg Kama, Judith Pallot, Brice Perombelon and Tim Schwanen for feedback and more general discussion on the themes of this book. I am especially indebted to the input and support I have received from my colleagues Linda McDowell and Richard Powell (now University of Cambridge). The seeds for the project were planted during my time at University College London, where in 2011 I completed my PhD, examined by Klaus Dodds (Royal Holloway) and Chris Browning (University of Warwick), and where I subsequently held a teaching position. At UCL I would in particular like to thank my former supervisor Felix Ciută for various forms of support and particularly for debate on the politics of war games. My thanks also go to Jason Dittmer (also UCL) for discussions on geopolitical architecture (and for co-editing the book series *Geopolitical Bodies, Material Worlds* with me at Rowman and Littlefield) and to Alan Ingram (also UCL) for prompting me to think about the interface of geo- and biopolitics in the first place.

I am particularly grateful for a close reading of my third chapter to Christian Abrahamsson (Universitetet i Oslo), Lucian Ashworth (Memorial University of Newfoundland), Andrew Barry (University College London), Audra Mitchell (Wilfrid Laurier University) and David T. Murphy (Anderson University) at the 2016 ISRF workshop on 'New earth thinking', organised by Richard Powell at Girton College, Cambridge. I have hugely benefitted from discussion at the 2015 European International Studies Association (EISA) conference in Giardini-Naxos, the 2014 and 2016 annual international conference of the Royal Geographical Society (RGS-IBG) in London, the 2014 regional conference of the International Geographical Union (IGU), the 2014 and 2015 annual conventions of the International Studies Association (ISA) in Toronto and New Orleans as well as seminars and workshops in Birmingham, Fulda, London, Oxford, Potsdam, Uppsala and Sheffield. I have enjoyed the support of a number of very helpful archivists at the Federal Archives in Koblenz and the German Military Archives in Freiburg and would particularly like to mention Doris Hauschke at the Federal Commissioner for the Stasi Records in Berlin. I have also benefitted greatly from conversations with Jörg Diester at Dokumentionsstätte Regierungsbunker and am very happy to have received the right to print the two images of camp Bellersdorf from Bernd Donsbach at Traditionsverband Aartalkaserne. All other images used in the book are my own. Before I forget, I should probably mention my students at St John's and Jesus College (particularly Harry Gibbs), with whom I have had the chance to debate various aspects of the material covered in *Cryptic Concrete*, and I am indebted also to two anonymous referees for the comments and suggestions and to the editor of the RGS-IBG series at Wiley, Dave Featherstone (University of Glasgow), for his help and patience.[1] All translations from the German are my own.

Last but certainly not least, I am particularly grateful for the support I have received from Anna (Toropova) who has been an absolute star in ways, academic and non-academic, that are too many to list here. I would also like to thank my family, particularly Lizzie, Jost, Gerlinde and Klaus. This book is dedicated to my mother Linda (1953–2011).

## Endnote

1   The book draws at times on some of my previous publications. My first thoughts on the topic were published in *Environment and Planning D* (Klinke 2015). Chapter 6 is a redrafted version of a paper that came out in *Transactions of the Institute of British Geographers* (Klinke 2016). Elsewhere the book is in dialogue with a co-authored piece that appeared in *Geopolitics* (Klinke & Perombelon 2015) and a number of shorter pieces I published as op-eds online.

# Chapter One
# Of Blood and Soil

## The Death of German Geopolitics

A concern with the politics of earth first emerged in Germany in the late
nineteenth century as geographers like Friedrich Ratzel (1844–1904) and his
disciple Karl Haushofer (1869–1946) sought to examine the causal influence that
a nation's position, climate or access to natural resources had on its rise and fall.
In turning to the concepts and ideas popular in the natural sciences, 'German
geopolitics' became known for developing a political theory that naturalised both
the territorial configurations of global politics and the phenomenon of interstate war.
Whilst geopolitical traditions were mushrooming elsewhere, too, German geopolitics was particular in its understanding of the state as a political life form (an
organism) that tried to secure its survival by conquering and defending *Lebensraum*
(living space). In the sense of articulating a theory of the interplay of life and
earth, these ideas were biopolitical in as much as they were geopolitical. Indeed,
we owe the terms 'geopolitics' and 'biopolitics' to Rudolf Kjellén (1864–1922),
another of Ratzel's followers (Kjellén 1920: 94).

Much like Halford Mackinder's perhaps more widely known brand of 'British
geopolitics' (see in particular Kearns 2009; Mackinder 1904), German geopolitics was mesmerised by what it saw as the eternal struggle between land and sea
power. Indeed, Haushofer argued for an alliance of the land powers of Germany
and Russia to counter what he saw as the predominance of Anglo-American naval
power. Haushofer and his contemporaries furthermore displayed a Malthusian

*Cryptic Concrete: A Subterranean Journey Into Cold War Germany*, First Edition. Ian Klinke.
© 2018 John Wiley & Sons Ltd. Published 2018 by John Wiley & Sons Ltd.

concern with overpopulation and desired for Germany to break out of its unfavourable centric position in Europe and establish a greater pan-Germany (Haushofer 1926[1979]: 532). Preoccupied with the idea of autarky, German geopolitics also encouraged an economy that would be self-sufficient rather than relying on trade with other nations. For Kjellén, the nation-state should, 'if necessary', be able to survive autonomously – 'behind closed doors' (Kjellén 1917: 162).

The proponents of German geopolitics would soon make a name for themselves by promoting their view of states as organisms that could grow and shrink and that were rooted in the territorial 'soil' on which they stood. In dialogue with the Darwinian ideas of his time, Ratzel wrote about state behaviour as a relentless struggle for 'survival' or 'being' (*Kampf ums Dasein*), which he saw in fact as a struggle for space (Ratzel 1901). Germany, the geopoliticians felt, needed to expand its living space in order to survive. This led to an emphasis on territorial space as a marker of a state's well-being. Indeed, Rudolf Kjellén thought it less problematic for a state to experience population loss than territorial loss (Kjellén 1917: 57). Finally, geopolitics was obsessed with questions of death, extinction and ruination, leitmotifs that emerged both from a social Darwinian preoccupation with survival and extinction as well as from a cyclical understanding of history. A key medium for the proliferation of geopolitical ideas during the interwar period and into World War II was the *Zeitschrift für Geopolitik*, founded by Karl Haushofer and Kurt Vowinckel in 1924.

German geopolitics would perhaps have remained a footnote in world history had Karl Haushofer not in 1924 been introduced to the as yet comparatively unknown Austrian politician Adolf Hitler, who was then putting together his manifesto, *Mein Kampf*. Given the geopolitical tone of significant parts of *Mein Kampf*, Haushofer would later be credited in the United States as the mastermind behind Nazi foreign policy (Ó Tuathail 1996). This idea of Haushofer as an *éminence grise* whose fictitious *Institut für Geopolitik* guided Nazi foreign policy has since been shown to be a myth (Murphy 2014). In fact, Haushofer's theory had 'long been drowned out by Hitler's ever-escalating pace of diplomatic crises, war, and extermination' by the time of the attack on the Soviet Union (Herwig 1999: 236; see also Murphy 1997: 244). Although Nazi ideologues continued to publish on questions of *Lebensraum* well into the war (Daitz 1943), Haushofer and Ratzel's geopolitics was in fact not in accord with Nazi ideology because it displayed an ambivalence towards the Third Reich's biological theory of race (Bassin 1987a). Despite these crucial ideological differences and the fact that Haushofer had even fallen out of favour with the Nazis, he came under attack after the war and would subsequently commit suicide in 1946. His suicide note famously asked for him to be 'forgotten and forgotten' (Haushofer 1946).

The story of this 'Faustian deal' between German geopolitics and the Nazis (Barnes & Abrahamsson 2015: 64) remains a crucial episode within the history of the geographical discipline until this day.[1] This story of German geopolitics is always told in much the same way, namely as that of a temporally confined period that starts with the publication of Ratzel's *Politische Geographie* in 1897 and ends in 1946 with Haushofer's suicide (Agnew 2003; Dodds 2007; Dittmer & Sharp 2014; Mamadouh, 2005; Ó Tuathail 1996). This, of course, does not mean that existing histories of geopolitics end in 1946, for geopolitics had a central role to play during the Cold War, as many observers have recognised (Dalby 1988, 1990; Dodds 2003; Ó Tuathail 1996). And yet, the existing literature has tended to see geopolitics as having managed to survive World War II precisely by renouncing its specifically *German* variant. In this reading, Cold War geopolitics, as it was articulated in the United States and elsewhere, is deemed to lack the biopolitical underside of a Ratzelian or Haushoferian geopolitics (Werber 2014: 143). This, as we will see, is certainly not the case.

Even in the work of those who have explicitly sought to explore the remains of German geopolitics after 1945, geopolitics is overwhelmingly seen to have been silenced in both the Federal Republic of Germany (FRG) and the German Democratic Republic (GDR). In this vein, postwar German Geographers like Troll (1949: 135) and Boesler (1983: 44) would speak self-evidentially of German geopolitics as having 'collapsed' (see also Michel 2016: 137). Others would suggest that the subdiscipline of political geography was 'marginalised' (Kost 1988: 2), whilst the wider discourses of geopolitics were 'stigmatised' (Kost 1989: 369). More recently, the historian Karl Schlögel (2011: 12) has bemoaned 'a lost tradition' of spatial thinking in Germany with the geographer Paul Reuber (2012: 90) writing of 'decades of complete silence' from geopolitical voices in German-speaking academia. Others have gone so far as to suggest that the Federal Republic 'civilianised' its geopolitics after 1945, swapping jackboots for Birkenstock sandals (Bachmann 2009).[2] We can find echoes of this kind of thinking in the idea that in the early twenty-first century the so-called 'German question' has only re-emerged in geo-economic rather than a geopolitical form (Kundnani 2014) or in the crude insistence that Germany is a 'quasi-pacifist' power (Kaplan 2012: 11).

This idea of a 'silence' around geopolitical concepts and ideas would seem to imply that the Germans heeded to Karl Haushofer's desire to disappear intellectually. In this way, observers have described his political testament as a 'funereal summing up of the demise of German geopolitics following the Nazi defeat in World War II' (Giles 1990: 13). But here we should not necessarily do Haushofer – or indeed his enemies – the favour of assuming that his ideas, and those of German geopolitics more generally, did indeed disappear alongside him. For although political geography *was* limited to the fringes of the geographical discipline in Germany and the terminology of geopolitics was indeed taboo in mainstream politics, this should not be read as evidence that *all* geopolitical writing was tabooed in Germany. Indeed, Bach and Peters (2002: 1) have remarked in passing that geopolitics 'never ceased to be a factor in German politics'. As Sprengel noted as early as 1996, however, such claims remain to be investigated in detail (Sprengel 1996: 36).

This book, then, sets out to establish what happened to this politics of life and earth *after* 1945. It argues that the twentieth century witnessed a second attempt to put a programme of geo- and biopolitics into practice in Germany, for an obsession with questions of national survival and space re-emerged through the equally geopolitical project of the Cold War in which the Bonn Republic, as the larger of the Third Reich's successor states, participated enthusiastically in the 1950s and 1960s. Like the Third Reich, the Cold War too expressed its logic of survival and extermination in material form. This resurfacing, or rather sub-mersion, of geo- and biopolitics in architectural form is puzzling given the seem-ing stigmatisation of geo- and biopolitics in post-1945 Germany. Nazi imperialism and the medicalising logic of extermination was seen, first by the Allies but increasingly by the majority of Germans too, as the root cause of a catastrophe that was not merely national but global in scale.

If we wish to unpack the seeming paradox as to why a particular amalgamation of geo- and biopolitics managed to survive its own funeral, I argue that we need to take seriously not just the level of intellectual discourse, where Ratzelian and Haushoferian concepts would cautiously re-emerge during the 1950s, but also pay attention to the militarised landscapes and subterranean spaces that the Cold War has left us with. For it is precisely here that we can get a sense of how this obsession with spaces of national survival managed itself to survive. As I will argue below, the Cold War turned the Nazi fantasy of the conquest for new *Lebensraum* into a more modest, though nevertheless bio- and geopolitical, fan-tasy of finding an *Überlebensraum* – a space of survival.

## West Germany and the Bomb

After the demise of the Third Reich, Germany disappeared from the map of European politics as a major power. Given the extent of the German war crimes and the scale of its defeat, its two successor states had to tread carefully in the international arena. Founded in 1949, only three years after Karl Haushofer's death, both the Federal Republic of Germany and the German Democratic Republic rejected not just the Third Reich's politics of extermination but also the tradition of German geopolitical thought that was now seen to have been the driving force behind Nazi expansionism. Indeed, the term geopolitics was rarely used in West German mainstream political discourse in the decades after the war, other than to discredit an opponent.

Although geopolitics was thus tabooed as a mode of thought in West Germany, it is important not to be deceived here, for geopolitics, as a discourse on global power struggles, was still very much alive. Bound to its Western allies, the young West German state swiftly adopted a rather radical version of Cold War geopoli-tics. From the early 1950s onwards, anti-Soviet sentiments would return to the fore, promoted by former Wehrmacht generals and the young republic's political

**Figure 1.1**   Konrad Adenauer on a visit to the United States (1961). Source: Das Bundesarchiv. Reproduced with permission.

elites. Some of these new geopoliticians would soon become advisors to the new German army, the Bundeswehr, their ideas chiming well with Konrad Adenauer, the Federal Republic's first chancellor (1949–1963). A devoted anti-Communist, Adenauer (Figure 1.1) soon became known for his hardnosed 'policy of strength' vis-à-vis the Soviet Union (*Politik der Stärke*) and the unambiguous 'Western orientation' (*Westbindung*) of German foreign policy, both of which formed important parts of the country's dominant foreign political narrative until the 1970s – and arguably after that, too.

Whilst Adenauer made sure to avoid the now discredited terminology of *Lebensraum*, his worldview was nevertheless geopolitical. In 1954, the same year that *Time* magazine made him 'man of the year', *Life* ran a special issue on the new West German ally, in which Adenauer was invited to lay out his vision. In true geopolitical manner, he explained to his American audience that 'an *Atlas of World History*' showed 'much more directly' than did written history that the 'area of Europe-Asia landmass in which freedom still prevail[ed]' had become 'frighteningly small in Europe since Russia's power reached the Elbe' (Adenauer 1954: 26). For Adenauer, the river Elbe, which served as a segment of the iron curtain, was

**Figure 1.2**   Conservative election poster: *'No – therefore CDU'* (1949). Source: © Konrad-Adenauer-Stiftung, 2017. Reproduced with permission.

no less than *the* boundary between Western civilisation and Eastern barbarism. In 1946, he famously warned that 'Asia was on the river Elbe', a statement that tapped into familiar racial representations of the Soviet Union (Adenauer 1946, see also Figure 1.2). Adenauer argued that the Soviet Union would simply 'overrun the rest of Europe' if it were to advance to the river Rhine (Adenauer 1949) and feared that 'the aggressive imperialism of Soviet Russia' would drive the United States out of Western Europe (Adenauer 1951). West Germany's ideological compatibility with the Western Allies' struggle with the USSR and its 'strategic' position in Central Europe was what permitted Adenauer's Federal Republic to join the European Coal and Steel Community in 1951 and the North Atlantic Treaty Organisation (NATO) in 1955.

Whilst West Germany was integrating into a military alliance that was actively preparing for a war with the Soviet Union, it was of course also interested in preventing the outbreak of such a war. A Third World War was likely to turn nuclear and thereby threaten to devastate the country for a second time in the space of just a few years. As a semi-sovereign state and a frontline territory, West Germany's strategic choices were of course limited. But rather than experimenting with a policy of neutrality, successive West German governments were convinced that they could only secure the country's independence from the Soviet Union by embracing the North Atlantic alliance and its policy of nuclear deterrence.

Indeed, Adenauer repeatedly rejected neutrality in the Cold War, claiming that 'the winds would blow radioactive clouds even over a Germany [...] that had declared itself to be neutral' (Adenauer 1957). The administration in Bonn, the new West German capital, was thus broadly in favour of NATO's nuclear arms buildup, for it felt that the irrationality of a nuclear war was the only chance to prevent a conventional war 'on German soil'. Thus, after joining NATO in 1955, Bonn tried to lobby the alliance for a policy of hardline nuclear deterrence. Indeed, the young semi-sovereign West German state had been keen to develop its own nuclear weapons programme, but had been forced to renounce any ambitions of acquiring atomic, biological or chemical weapons at an early stage. Too controversial was the idea that Bonn would accomplish what the Third Reich had failed to achieve. And yet, NATO soon worked out a compromise, the so-called nuclear sharing initiative, which allowed non-nuclear powers like West Germany to participate in nuclear planning, stationing and delivery of nuclear weapons. As a consequence, the West Germans could at least simulate some form of nuclear status.

In a sense, however, this policy only allowed Bonn participation in what had already begun on the ground, namely the arming of West German territory with US battlefield or tactical nuclear weapons. Tactical nuclear weapons are small-yield and short-range nuclear missiles and bombs that are primarily designed for use in battle. They are usually distinguished both from high-yield *strategic* nuclear missiles that are launched initially from aircraft and later from missile silos and nuclear submarines that are meant to take out entire cities or large military installations. Whilst tactical nuclear weapons were meant to make up for Western Europe's perceived conventional military weakness vis-à-vis the Red Army, their 'military value could never be properly explained' (Freedman 2013: 171). Given that tactical nuclear weapons were often mobile, they were much less exposed to enemy attack than stationary missile launch silos. This meant that they were something of an unknown quantity that could escalate a conventional military war and transform it into an all-out nuclear Armageddon.

Starting in the early 1950s, the nuclearisation of West Germany had met little resistance from the leadership in Bonn, but in order to build a consensus amongst the population the government had to euphemise these new weapons. And so, in 1957, Chancellor Adenauer would declare that tactical nuclear weapons were 'nothing but the further development of the artillery' (Der Spiegel 1957, see Figure 1.3). As historians started to point out by the late 1980s, West Germany's frontline soldiers had in the 1950s been assigned 'the task of staving off a Soviet assault long enough for NATO to drop nuclear warheads above them', rendering them the 'atomic cannon fodder of a future war' (Cioc 1988: 9). In the early

**Figure 1.3**    Tactical nuclear weapons paraded on the Nürburgring (1969). Source: Das Bundesarchiv. Reproduced with permission.

1980s, the growing anti-nuclear movement would argue that NATO's first use doctrine would destroy what it meant to defend and therefore constituted a 'suicidal form of defence' (Afheldt 1983: 13).

This suicidal politics was not unique to West Germany. As the social theorist Paul Virilio has argued more generally about the relationship between modern war and the state,

> As destruction became a form of production, war expanded, not only to the limits of space but to all of reality. The conflict had become limitless and therefore endless. It would not come to an end, and, in 1945, the atomic situation would perpetuate it: the state had become suicidal (Virilio 1975: 58).

If we wish to understand this paradoxically suicidal politics that emerged in Cold War West Germany in crystallised form – but which was more generally characteristic of the Cold War – then we need to read Virilio's statement in conjunction with contemporary debates around biopolitics. As Michel Foucault would come to argue, '[i]f genocide is indeed the dream of modern powers, this is not because of a recent return of the ancient right to kill, it is because power is situated and exercised at the level of life, the species, the race, and the large-scale phenomena of population' (Foucault 1978: 137).

## Towards a Cold War Biopolitics

Fueled primarily by its prominent role in theorising the global war on terror, biopolitics has now established itself as an important focal point for debate across human geography and has inspired scholarship on topics as diverse as migration, HIV, airports, climate change and food provision (for an overview see Rutherford & Rutherford 2013). The concept of biopolitics tries to capture how the production and protection of life is articulated with the proliferation of death. Biopolitics, in other words, is the moment when politics is applied to the boundary between the natural and the social, indeed to life itself (Lemke 2011). In political geography in particular, notions of the biopolitical have been drawn on to examine the re-emergence of the camp in the war on terror and the US-led occupation of Iraq (Diken & Lautsen 2006; Gregory 2007; Minca 2005, 2006, 2015), the Arab–Israeli conflict (Ramadan 2009) and in post-9/11 border regimes (Amoore 2006; Sparke 2006; Vaughan-Williams 2009, 2015). In doing so, political geographers have made a powerful connection between Nazi biopolitics, on the one hand, and the contemporary security state, on the other.

Geographers have tended to approach biopolitics through the works of two social theorists, Michel Foucault and Giorgio Agamben.[3] Foucault began in the late 1970s to trace the ways in which life itself had increasingly become the object of modern governmental practices. He famously argued that the sovereign's old right over life and death had increasingly given way to a new power, a power of fostering life, even to the point of disallowing it. In his words, 'the right to take life or let live' was replaced by 'the right to make live and let die' (Foucault 1976: 241, 1978: 137). Rather than seeing biopolitics as a response and remedy to violence and war, Foucault saw them as compatible. The death of inferior forms of life made the life of the population healthier. Controversially, he argued that a form of power that combined the right to kill and the exposure to death 'was inscribed in the workings of all states' (Foucault 1976: 260). In an often quoted passage, Foucault argues that

> Wars are no longer waged in the name of a sovereign who must be defended; they are waged on behalf of the existence of everyone; entire populations are mobilised for the purpose of wholesale slaughter in the name of life necessity: massacres have become vital. It is as managers of life and survival, of bodies and race, that so many regimes have been able to wage so many wars, causing so many to be killed (1978: 137).

And yet, what often gets missed when this passage is taken out of context is that Foucault was in fact speaking about nuclear war. For he continues,

> through a turn that closes the circle, as the technology of wars has caused them to tend increasingly toward all-out destruction, the decision that initiates them and the one that terminates them are in fact increasingly informed by the naked question of

survival. The atomic situation is now at the end point of this process: the power to expose a whole population to death is the underside of the power to guarantee an individual's continued existence (ibid).

Even nuclear war, Foucault held, was thus fought not in the name of the sovereign but legitimated through the preservation of the population's life – paradoxically to the point of risking its very death (Foucault 1976: 253).

More recently, Giorgio Agamben has built on these pioneering insights but departing from Foucault by refocusing biopolitics on the question of sovereignty and the law. Zooming in on the simultaneously legal and extra-legal logic of the state of exception, a conceptual focus he takes from Carl Schmitt (1922[2005]), he has argued that sovereign power constitutes itself and its counterpart, 'bare life', in and through exceptional spaces. In a now widely familiar argument, Agamben traces this production of bare life back to a distinction in Ancient Greek between political (*bios*) and natural life (*zoe*). He claims that biopolitics functions as the calculation, government and abandonment of *zoe*, though in ways that ultimately include it through its very exclusion. He finds this rendering of bare life again in the Roman legal figure of *homo sacer*, an expellee from the community who cannot be meaningfully sacrificed and can therefore be killed with impunity (Agamben 1998: 102). *Homo sacer* is human existence that has been deprived of its rights and its political voice – life that is exposed to death. Through a process of progressive normalisation, Agamben argues, states of exception have a tendency to turn democracies into dictatorships and render their citizens potential *homines sacri* (ibid.: 111).

Agamben has held that these logics were most visible in the Third Reich's rule of emergency decrees and more specifically in the tightly sealed and hygienic space of the concentration and death camp. As Minca (2005: 407) following Agamben puts it, sovereign power requires the camp as 'a material and mappable space within which violence becomes the constitutive element of *both* the torturer and the victim'. In this reading, the Nazi camps must be grasped as intimately intertwined with the fantasy of an Eastern empire. 'The "shrinking" of the internal spaces of the camp', Giaccaria and Minca (2011a: 5) have held, 'was a functional and symbolic counter-dimension to the expansive nature of German *lebensraum*'. The vast and open living space, in other words, relied on the cramped and concentrated space of the camp as a means of racial 'purification'. Like others before him (Bauman 1989), Agamben suggests that the deadly dimension of biopolitics (thanatopolitcs) as materialised in Auschwitz was not so much a flaw in the project of modernity as its constitutive and repressed underside. The camp, in other words, was and remains today the hidden paradigm of Western modernity (Agamben 1998: 181).

This politically pessimistic reading of Western modernity has not been without its critics. Ernesto Laclau has accused Agamben of a 'naïve teleologism' that fails to see the messiness of modern biopolitics and precludes the emergence of modernity's emancipatory potential (Laclau 2007: 22). Pushing this critique further, Mark Mazower has questioned whether Auschwitz should be read as *the* symbol of the genocidal episode of World War II. He points especially to the death of 2.1 million Soviet soldiers in German POW camps during 1941/42 that occurred as part of a logistic failure rather than the bureaucratic and medicalising logic of biopolitics foregrounded by Agamben (Mazower 2008: 31). This latter point highlights an important omission in both Foucault's and Agamben's original theorisation of biopolitics, namely the question of geopolitics, military logistics and the embeddedness of the Holocaust in a wider landscape of total war.[4] This absence is particularly surprising given that the term 'biopolitics' originates in the work of an aforementioned geopolitician, the Swede Rudolf Kjellén.

By focusing its analysis on the war on terror and the geographies of the Third Reich, the literature on biopolitics has moreover, with few exceptions (Collier & Lakoff 2015; Monteyne 2011),[5] overlooked the Cold War as a biopolitical set of events. Whilst geographers have long argued that Cold War *geo*politics emerged smoothly from Nazi (Ó Tuathail 1996: 87) and imperial geopolitics (MacDonald 2006b), the relationship between Nazi and Cold War *bio*politics is yet to be systematically explored. This is perhaps somewhat surprising given that the Cold War was a formative period for political geography. Indeed, the last years of the Cold War served as the political milieu for the emergence of critical geopolitics, the body of literature that sets out to examine the persistence and political impact of geopolitical discourse (Dalby 1988, 1990a, 1990b; Ó Tuathail & Agnew 1992; Sharp 1993).

I argue that if we want to understand better the intersection of geo- and biopolitics, we must return to the works of those geographers who first articulated a theory of *Lebensraum*. The role that the tradition of German geopolitics played within the development of biopolitics is often noted (Esposito 2008: 16; Lemke 2011: 13), but rarely expanded on. When Kjellén first coined the concept of biopolitics in 1920, he expressed his desire to capture both the physical (read natural) and cultural (read social) dimensions of life (Kjellén 1920: 94). Geopolitics, on the other hand, for him was the 'doctrine of the state as a geographical organism or as a phenomenon in space' (Kjellén 1917: 46). From their conceptual origins, the *geo* and the *bio* were thus difficult to disentangle. Haushofer, for instance, would refer in the 1930s to the phrase 'blood and soil' as 'an inextricable community' (Karl Haushofer 1935: 11). The phrase would play a crucial role in defining the German nation in territorial and racial terms under National Socialism in ways that would enable the extermination of those ethnicities that lacked a 'grip' on the soil. Following up on Roberto Esposito's claim that '[n]azism has much in common biopolitically with other modern regimes' (Esposito 2008: 111), it becomes necessary to grab biopolitics by its intellectual roots.

Geopolitics and biopolitics thus need to be considered together as two iconic forms of power that have operated at the very heart of modern statehood.

There is 'no geopolitics that does not imply a correlate biopolitics, and no biopolitics without its corresponding geopolitics', Dillon and Lobo-Guerrero (2008: 276) have us know. Stuart Elden adds that 'geopolitics works with similar operative principles' to biopolitics, namely calculation and metrics, 'both categories emerged at a similar historical juncture as new ways of rendering, understanding and governing the people and land' (Elden 2013: 49). Indeed, both geo- and biopolitics have a dark history: whilst we can find within the seemingly protective practice of biopolitics a much less benign policy of excluding forms of life that are deemed dangerous and unworthy, geopolitics has historically been used to legitimate territorial expansion and colonial oppression. As Giaccaria and Minca (2016: 3) argue, it is 'difficult if not impossible to operate a distinction between life and space, between bio-politics and geopolitics, since the Third Reich incorporated *Lebensraum* by merging its duplicitous meaning, as living/vital space and as life-world'. Yet, whilst it most certainly makes sense to think geo- and biopolitics together, and it is of course the purpose of this book to do so, it is necessary to be aware of a conceptual tension between the two in the existing literature, a tension which this study attempts to bridge. Whereas geopolitics is often treated as a *body of knowledge* that is concerned with the mapping of territories, biopolitics is represented as a *mode of government* which targets the life of the population, fostering and abandoning it. More often than not, geopolitics is thus approached as a foreign political discourse and biopolitics as a practice of domestic politics.

Instead, geo- and biopolitics should be seen as dealing with *both* the domestic and the international realm and constituting *both* a body of knowledge on the state as organism (the tradition of German geopolitics) *and* a particular mode of governance (such as the German occupation of Eastern Europe during World War II and the nuclear politics of the Cold War). Moreover, I argue that we need to make a second important amendment to recent debates on biopolitics. We need to dethrone the camp as the most important space in which geo- and bio-politics intersect. For unless one assumes the concentration camp to be a mere metaphor for the politics of modernity, it is difficult to read it as the defining space of the Cold War. Of course, the Cold War's many military conflicts, from Korea to Afghanistan, involved POW camps, but the Cold War did not simply see the death camp or indeed the fantasy of *Lebensraum* re-emerge in the same way in which they had played a role under National Socialism. Yet, a close look at what was arguably the most iconic of Cold War spaces will reveal a stealthy morphing and inverting, a moving underground and a pouring into concrete of the logic of *Lebensraum*. In order to grasp the latter, it is necessary to grapple with a very particular form of passive aggressive architecture and a biopolitical space in its own right, the bunker. In doing so, we will reveal the return of the extermination camp, though this time as a space from which the production of the corpse has been outsourced.

## The Bunker and the Camp

Whilst the bunker has its origins in the context of World War I trench warfare, it came to the fore in the course of the technological possibility and political will to take out whole cities in World War II. Crucial to any understanding of the bunker's spatial politics is the work of Paul Virilio and his conception of the bunker as a spectacular, cryptic and insecure monolith. Captivated by the aesthetics of Hitler's Atlantic wall, his seminal *Bunker Archaeology* (1975) approaches the bunker as a monolithic space that promises survival in an era of total war, an age in which weapons have become so omnipotent that distance can no longer act protectively. Virilio discusses the bunker as a paradoxically secure and insecure space. Although the Atlantic wall was the harbinger of a new age of total war (1975: 45), it was also as a symbol of weakness and imperial overstretch, a strategic still birth, as it were. But whereas for Virilio, the bunker is always theatrical, built for an external gaze from which it would seem impenetrable (ibid.: 47), its actual value is often psychological rather than strategic; it tries to forge an identity where such an identity is in fact under threat.[6] We have already learned about Virilio's insistence that bunkers are cryptic in the sense of being places of shelter, worship and salvation.

It is specifically the subterranean celebration of death in the crypt that links the nuclear bunker to the biopolitical space of the camp. Indeed, as a subterranean survival capsule, the bunker emerges in the same political context as the death camp, chosen by Agamben as a symbol for modernity's underside – total war. As Esposito reminds us, it is only in such a war that one can kill 'with a therapeutic aim in mind, namely, the vital salvation of one's own people' (Esposito 2008: 136). It is precisely this logic of survival and extermination, which draws together social Darwinism, German geopolitics, total war and the complementary architectural spaces of the bunker and the camp.

In developing this argument, it is useful to follow Benjamin Bratton, who, in a 2006 foreword to Virilio's *Speed and Politics*, briefly stages an encounter between Virilio's writings on the bunker and Agamben's reading of the camp to argue that both spaces should be understood as 'doubles'. He notes that both are 'hygienic' and 'defensive' but whereas the bunker functions as 'a concrete prophylactic, the camp is incarcerating' (Bratton 2006: 19). Indeed, both spaces attempt a radical inside/outside distinction. Whilst one 'is an architectural membrane against a hostile world', the other performs 'an expulsion-by-enclosure of the Other from the normal performance of law' (ibid.). In fact, however, the nuclear bunker turns the biopolitical logic of the camp inside out. Whereas the Nazi empire's Eastern living space was punctuated by concentration and death camps, the subterranean living spaces of the nuclear bunker would be scattered across a potential landscape of Cold War extermination. Following Virilio, Bratton goes on to emphasise the role of logistics as the fundamental socio-technical condition of possibility for both spaces, 'where the only compulsion is the execution of governance on a raw mass, mobilising it, diagramming it' (ibid.).

Any exploration of the similarities and dissimilarities between the nuclear bunker and the concentration camp will have to start by noting the obvious, namely that the Cold War, unlike World War II, did not fully unleash its logic of extermination. Bratton is right to identify the logistics of governance as being at the heart of the two spaces but his insistence that they are 'often architecturally identical' is more questionable. After all, it is difficult to see how the nuclear bunker's reinforced concrete would resemble the notorious huts, fences, watch-towers and crematoria of Auschwitz. Moreover, it seems problematic to treat the nuclear bunker as a passive and static space, as Virilio does, since the bunker too is a place from which the Other is expelled. As I hope to illustrate below, Agamben's sovereign decision over life and death became localised in the nuclear bunker and materialised in its architecture. In important ways, the protective space of the nuclear bunker reproduced and inverted the lethal space of the camp.

Whilst the nuclear weapons storage site was clearly a space with a concentration camp-like architecture from which extermination could be rationalised and planned, it was ultimately a potential death camp from which the bodies had been purged. In this sense, the camp was designed to keep intruders out rather than to incarcerate and exterminate on site. Moreover, although few civilians came into contact with these deadly spaces and their exact location was of course hidden, their existence *per se* was not kept secret. Indeed, NATO's nuclear posture was publicly celebrated by the Cold War state and enjoyed considerable public support, too. Similarly, nuclear bunkers were not simply reproductions of the camp, but turned the latter's biopolitics inside out. They were built as subterranean colonies, securing their privileged inhabitants from the nuclear holocaust outside.

## Approach and Structure

By approaching the question of geo- and biopolitics both on the level of strategic discourse and on that of the built environment, this book hopes to contribute to a number of existing and emergent debates on the materiality of geopolitics. Whilst political geographers initially studied (Cold War) geopolitics by concentrating on elite and popular discourses under the banner of 'critical geopolitics' (Dalby 1990a, 1990b; Ó Tuathail & Agnew 1992; Ó Tuathail 1996; Sharp 1993, 2000b), the 2000s brought a wave of work that sought to venture beyond the mere textuality of geopolitics (Dittmer 2013; Müller 2008; Thrift 2000). Whilst feminist geographers began to draw attention to the corporeality and everyday fabric of geopolitics (Dowler & Sharp 2001; see also Dixon 2014), work around urban geopolitics increasingly focused on the construction and destruction of the built environment in armed conflicts (Fregonese 2009, 2012a; Graham 2004a, 2010, 2016; Weizman 2007). As geographers started to return to the questions of materiality which had been barred by a resistance to historical materialism and classical geopolitics, arguments about the political agency of the earth, from deserts (Squire 2015) and oceans (Steinberg & Peters 2015) to its geologic forces (Clark 2013a, 2013b), were increasingly heard.

The consequence of this re-materialisation of geopolitics has been a new concern with the 'enactment of bodies, things, and contexts that constitute the "landscape" of geopolitics' (Squire 2015: 148). It is 'to rethink our identities as agents of geological change, and in the process understand humanity's role in the larger order of things in new ways' (Dalby 2007: 112). As political geography has moved away from a concern with political boundaries and towards one with networks and fluids (Barry 2013a), political agency has been extended to non-humans (Dittmer 2013: 397). Although much of these debates have offered valuable additions and correctives to the textual preoccupations of a critical geopolitics, they have a tendency to lose the original subject out of sight, namely the intellectual tradition of geopolitics and the question of war and peace more broadly. There are of course arguments to be made as to why a narrow focus on the geopolitical tradition is problematic, but it seems as if much remains to be gained from an engagement with early twentieth-century geopolitical thought – for its conceptual categories may prove to be more resilient than is sometimes assumed.

Much of the debate on the materiality of geopolitics departs from the assumption that geopolitics is a flat discourse that neglects the vertical dimension of power, the way in which authority is administered from above (Elden 2013; Graham 2004, 2016). Similarly, it has been argued that the geographical abstractions inherent in geopolitics have shown a disinterest in the physical infrastructure that makes the state and modern warfare possible in the first place. These voices have taken their cue from Eyal Weizman who has held that geopolitics

> largely ignores the vertical dimension and tends to look across rather than to cut through the landscape. This was the cartographic imagination inherited from the military and political spatialities of the modern state. Since both politics and law understand place only in terms of the map and the plan, territorial claims marked on maps assume that claims are applicable simultaneously above them and below (Weizman 2002: 2).

Of course, it is true that geopolitics has tended to map the world in two-dimensional ways, either representing things that are in reality under the earth's surface or in its atmosphere cartographically or ignoring them altogether. And yet, as we will find out below, German Cold War strategists and civil defence planners had increasingly realised this problem by the 1950s by juxtaposing in their writings the geopolitical map and the architectural blueprint. It was precisely the impossibility of territorial conquest and the destructive power of nuclear war that led these new geopoliticians to seek *Lebensraum* in the third dimension.

Interestingly, Cold War research has also experienced its own 'material turn'. In the early 2000s, Rachel Woodward (2004: 5) had still rightly bemoaned that the material spaces of geopolitics and militarism still remained off limits for geographers (Woodward 2004: 5). Whilst historians and political scientists had long thought of *the architecture of the Cold War* as simply the balance of power, architectural historians had been obstructed by the fact that there is, as Vanderbilt (2002: 17) puts it, simply 'no Le Corbusier of the missile silo'. As the Cold War slowly slipped away as a key referent point in political debates and many sites were abandoned by the military, geographers and urban planners started to become more interested in the places where this conflict quite literally 'mattered' (Bartolini 2015; Bennett 2011a; Davis 2008; MacDonald 2006b; see also Woodward 2014: 46). This has come at a time when journalists and scholars across the social sciences have started to take a more active interest in the period's military landscapes, from atomic test ranges and listening stations to missile silos and deep shelters (Bennett 2017; Cocroft et al. 2005; Masco 2009; Monteyne 2011; Vanderbilt 2002; Wills 2001).[7]

Whilst this scholarship has offered us fascinating insights into the everyday politics and heritage of nuclear bunkers and missile silos, it must account for a number of lacunae, to which this book seeks to attend. Firstly, this literature has displayed an Anglo-American regional bias that has largely neglected the military landscapes of continental Europe. In this way, the bunkered Cold War society emerges as a reincarnation of the American frontier (Heefner 2012; Masco 2009) or of American consumer capitalism (Wills, forthcoming; see also Marcuse 1964[1999]: 93). Bunkers are seen to have been imagined as 'spacious, ridiculously well-stocked playrooms with artificial sunlight and state-of-the-art entertainment systems, inhabitable for years and years' (Vanderbilt 2002: 110–11) – but not as geo- and biopolitical spaces. Secondly, whilst this literature has drawn attention to the simultaneous exceptionality and everydayness of Cold War architecture, it has yet to systematically theorise notions of biopolitics. Moreover, with Cold War architecture frequently cast as an 'amusing' curiosity (Rose 2001: 12), its historical continuities with the aggressive architecture of National Socialism have so far gone unnoticed.

By re-materialising geopolitics without losing the intellectual tradition of geopolitics out of sight, I hope to show how ideas about the politics of life and earth were intimately entangled with the built environment during the Cold War. I seek to examine 'the repressed spaces of nuclear modernism; that is, the social logics, technoscientific practices, and institutional effects that were rendered invisible by this national fixation on extinction' (Masco 2006: 4). I am thus interested in the way in which discourses 'enter into construction and how in consequence buildings or planned environments become statements' (Hirst 2005: 158). In a Foucaultian vein, this offers 'a link between a discursive formation, the institutional conditions in which it becomes a practice, and the products of that practice' (ibid.). This is to recognise the ways in which soldiers and civilians are placed 'within symbols, not merely to write or speak about buildings in war, but to symbolise what mattered through the buildings themselves' (ibid.: 192). That said, my point is not to reduce the question of materiality to its semiotic dimension. I am not just interested in how materials were politically instrumentalised, but also in how materials such as poured concrete or razor wire had a life of their own and sometimes managed to resist the purposes to which they were put. We will see both how the problem of human waste enabled a legally exceptional geography of sewage disposal in the West German government's nuclear bunker and how the inability to hide the often gigantic sites of Cold War geopolitics attracted the anti-nuclear movement during the 1980s. And yet, I should clarify that I am not primarily interested in telling a story about material agency. There is little to suggest that the re-emergence of geo- and biopolitics will be materially produced. It is rather the entanglement of politics and technology, materials and ideas that this study sets out to explore. In making an argument about the relationship of discourse and materiality, this study develops three recurring themes: inversion, temporality and play.

Firstly, I argue that the geo- and biopolitical fantasies of national survival and extermination managed to survive the demise of fascism not just by going underground, where infrastructure was safe from blast waves and radiation, but by means of architectural inversion. Whilst the tactical nuclear missile camp sought to rationalise and enable the possibility of extermination, the nuclear bunker was designed to protect fragile human bodies from these very same forces of extermination. Whilst the former was designed as a space of self-extermination, the latter was to function as an inverted concentration camp, a space that protected its concentrated living space from the nuclear holocaust outside. Both spaces were effectively turned inside out and thereby obscured their architectural origins. As we will see below, we can find similar inversions of geopolitical concepts at a more intellectual level. A second returning theme through which the book seeks to link the discursive and the material is chronopolitics, the politics of time (Klinke 2013). I am particularly interested in teasing out the temporal relationships that are materialised in Cold War sites, such as the ideas of 'time capsule' and 'post-apocalypse' and the ways in which these concrete structures anticipated and tried to resist their own destruction. But I am also attuned to the ways in which a cyclical understanding of world politics that naturalises the rise and fall of nations found expression in the fantasy of the nuclear ruin.

A final theme through which I hope to link the Cold War's built environments and the strategic narratives that sustained it is ludic geopolitics, and the current engagement with and debates around it (Carter, Kirby & Woodyer 2015; MacDonald 2008). For nuclear bunkers and missile sites did not just exist in a dormant state, but were often used playfully in preparation for a war that never came. In this, the book wishes to attend to the importance of studying the 'everyday' articulations of geopolitics even in 'elite' settings (Dittmer & Gray 2010; Pain & Smith 2008; Thrift 2000). I focus in particular on a number of NATO war games that were performed from the 1960s until the 1980s, and argue that they brought out and played on the German trauma of urbicide and bunker life. In unpacking these ultimately rather *unplayful* games, I attempt to answer the question as to why West Germany, a state that was in many ways eager to construct itself as having broken with its national history, was in other ways very keen to repeat its historical mistakes. Why, to rephrase the question in a Freudian register, did the German elites compulsively repeat the trauma of geo- and biopolitical annihilation?

In terms of method, the book draws on a combination of archival research and intellectual history, as well as media analysis and web research. The bulk of archival materials are made up of declassified files from the 1950s and 1960s located in the Federal Archives in Koblenz and the German Military Archives in Freiburg. I have also drawn on a number of online archives, such as that of the German parliament and the Federal Office of Civil Protection and Disaster Assistance. Given the Cold War state's secrecy, the analysis is necessarily limited to what the Federal Republic has been willing to open up to the public. As many of the files on atomic weapons and nuclear bunkers were shredded at the time or remain classified today, my research also draws on East German intelligence located in the Federal Commissioner for the Stasi Records in Berlin and at the Military Archives in Freiburg. The analysis also offers a qualitative analysis of key West and East German daily and weekly newspapers and magazines, as well as a variety of informal publications, such as the pamphlets produced by the anti-nuclear movement. Finally, the book incorporates photographic documentation taken during fieldtrips to the sites at hand. Throughout, these materials are jux-taposed with historical photographs, mainly found in the Federal Archives. The point of this exercise is not just to make the Cold War landscapes come to life, as it were, but to suggest that the architectural forms that are produced in these nuclear landscapes invoke previous attempts to organise a society around the principles of survival and extermination.

The book is divided into seven chapters. This introductory chapter has set out the central argument; next up, a contextual chapter explores the origins of the intersection of geo- and biopolitics, both at the level of intellectual discourse and on that of material architecture before 1945. Four empirical chapters will then discuss, in turn, the renaissance of geopolitical thought in the 1950s, the birth of the nuclear bunker, the emergence of the tactical nuclear weapons camp, and finally the practice of nuclear war gaming. The concluding chapter outlines the study's wider theoretical and political implications and poses questions about the nature of the Cold War's geopolitical ruins.

Chapter 2 lays both the historical and conceptual basis for the discussion that informs subsequent chapters. I begin by unearthing a fascination with political earth and political life within the late nineteenth- and early twentieth-century tradition of German geopolitics. In doing so, I return to the foundational works of Friedrich Ratzel, Rudolf Kjellén, Karl Haushofer and others associated with German geopolitics in an attempt to show how these theorists of *Lebensraum* were not just fascinated by the nation's vitality and growth, but also by its death, extinction and extermination. I use this to argue that we can find within the idea of *Lebensraum* the seed of necropolitics (the politics of death), understood both as an aesthetisation of death and as the intellectual basis for a policy of survival and extermination. In the second part of the chapter, I take a closer look at the first attempt to put such a programme of national survival and extermination into practice, the Third Reich's conquest of *Lebensraum* in Eastern Europe. I discuss in particular the bunker and the camp as two archetypical material spaces through which this project was to be realised and I use this discussion to suggest ways in which we might begin to understand the Cold War in similar terms.

Chapter 3 seeks to establish what happened to the politics of earth in Germany after 1945. It argues that although the terminology of *Geopolitik* was largely banned from the public realm in the 1950s and 60s, the preoccupation with territory, national survival, autarky and living space managed to re-emerge in the preparation for World War III. This return of geopolitics may have been spawned in the pages of the reborn journal *Zeitschrift für Geopolitik*, but it was crucially amongst military strategists and civil defence planners rather than in academic circles that ideas about *Raum* (space) found a new platform. By surveying geopolitical writing in the first decades after the war, I examine the work of a number of 'intellectuals of statecraft' (Ó Tuathail & Agnew 1992), focusing on a number of figures who have so far not been examined as geopolitical thinkers in their own right. I discover not just that geopolitical ideas continued to be published in West Germany, but that they were moreover promoted by key political agents who were tasked with the establishment of the new West German army in 1955 and with the preparation for nuclear war. As Detlef Bald (2005) and Alaric Searle (2003) have shown, men like Heinz Guderian, Friedrich Ruge and Erich Hampe had a crucial influence on the strategic orientation of the new army and therefore the young republic. Much like the interwar geopoliticians, these men displayed a concern with the land and sea dynamic and with Germany's unfavourable position at the centre of Europe. Germany's territorial borders remained a concern, even though the aim of territorial expansion was abandoned. Interestingly, the new German geopoliticians continued to read the state as an organism that struggled for living space. But now living space was no longer understood as territorial but as architectural. In order to ensure Germany's survival in the era of atomic warfare, living space had to be sought in subterranea.

Chapter 4 picks up from the previous one by exploring 'in depth' the nuclear bunker, a space which materialised like few others the idea of geopolitical autarky. The chapter starts by examining the status of the bunker within West Germany's civil defence programme. It then zooms in on the uncanny story of a disused railway tunnel that was transformed into a slave labour camp during the early 1940s, only to be reborn in the late 1950s as the West German governmental nuclear bunker. In an attempt to dissect this bunker, the chapter explores the political and legal context in which the bunker emerged, its technical and security features and the organisation of everyday life underground. Moving from a discussion of the Federal Republic's emergency laws to the bunker's decontamination facilities, the chapter argues that the bunker's drive to protect its concentrated living space from the nuclear holocaust outside turned the concentration camp's biopolitical logic inside out. Whilst the governmental nuclear bunker was built to protect lives in a conflict that threatened to rid the earth of life itself, it offered very little protection for the population. Rather, it served as a mausoleum for the Cold War state.

Chapter 5 moves from a space of protection to a space of extermination. Examining West Germany's active participation in nuclear war, it starts by unpacking Bonn's participation in NATO's 'nuclear sharing' initiative. It then goes on to examine the public and bureaucratic debates around the establishment and maintenance of tactical nuclear weapons systems, the range of which was so short that they could only be used on West German soil. In the event of a nuclear war with the Warsaw Pact, these weapons would have been used to sacrifice large parts of the West German population, a biopolitical logic of 'immunisation' *par excellence* (Esposito 2008). This chapter is interested both in the ways in which these sites were legitimated and how they functioned materially. It explores how the nuclear weapon site replicated the spatial logic of the death camp – that iconic architectural space that stood at the intersection of Nazi geo- and biopolitics. The missile camp too was an ultimate 'space of exception' within which extermination was rendered possible and even rational. It was enabled and governed by a highly concentrated but sometimes very mundane form of sovereign power, which was obsessively committed to the creation and fortification of hermetically sealed spaces. I conclude that the tactical missile camp stretched Esposito's biopolitical paradigm of immunisation to its vanishing point – national self-destruction and potentially the destruction of life on earth.

In an attempt to bring the nuclear weapons storage site and the governmental nuclear bunker into a dialogue, Chapter 6 takes as its focus a string of nuclear war games that the West German government staged in its command bunker between 1966 and 1989. After a discussion of civil defence exercises and war games in West Germany more generally, the chapter looks in more detail at Fallex 66, a 1966 NATO game. During this exercise, the Bonn Republic controversially simulated nuclear strikes on 'own' West German targets and a resupply of forces after a nuclear war on German territory. Whilst in line with NATO nuclear strategy at the time, the self-destructive simulation was read in East Berlin as an outburst of excessive and obscene enjoyment. Drawing on a psychoanalytic reading, the chapter argues that Fallex 66 and similar exercises should be understood not as a mere enactment of Cold War geopolitics, but as a Freudian 'fort-da' game, a traumatic re-enactment that was tellingly set in the subterranean space of a German bunker.

# Endnotes

1  Whilst earlier debates focused in particular on the work and influence of Friedrich Ratzel and Karl Haushofer (Bassin 1987a, 1987b; Heske 1986, 1987; Kost 1987; Smith 1980), more recent discussions have focused on the role of Carl Schmitt and his geopolitical writings (Barnes & Minca 2013; Elden 2010; Legg 2011; Minca & Rowan 2014; 2015a; 2015b).

2  It is based precisely on these kinds of arguments that debates emerged in the 2000s that spoke of a 'renaissance' of geopolitics in Germany (Bassin 2003; Behnke 2006).

3  We may want to add Michael Hardt and Antonio Negri's 2000 book *Empire* here, which was key to the alter-globalisation struggles of the 2000s. Here biopolitics denotes a new stage of capitalism, one characterised by the dissolution of the boundary between economics and politics, production and reproduction.

4  For a discussion of the silences around colonial violence in Agamben see the edited collection by Svirsky and Bignall (2012), especially Atkinson's chapter on Italian imperial aggression in what is now Libya.

5  Monteyne (2011) has possibly produced the most comprehensive existing study on nuclear bunkers. Located in the discipline of architectural history, it concentrates on civil defence but only mentions biopower in passing, without seeking to tease out the connections to previous the Third Reich's politics of survival/extermination. Hornblum et al. (2013) expose Cold War era medical experiments on children that were designed and carried out by an alliance of US scientists, pharmaceutical corporations and the US military. In doing so, they come closest to producing a study of Cold War biopolitics that could link up with the medicalising logic of biopolitics offered by the Nazis, although it is not conceptualised by them as such.

6  Whilst Virilio's prolific writings have inspired generations of thinkers, they have to be approached with the necessary critical distance for they are themselves caught up both in the geopolitical gaze (Luke & Ó Tuathail 2000: 364) and bunkered thought (Gane 1999: 100). Moreover, Virilio's insistence that bunkers are monolithic clearly has its limitations when different layers of geopolitical sedimentation in the biographies of specific bunkers are taken into consideration (Hirst 2005: 213).

7  For systematic attempts to rethink the urban geopolitics of the Cold War, see Farish and Monteyne (2015).

# Chapter Two
## *Lebensraum* and Its Underside

Few of those who utter the phrase "blood and soil" are aware that the modest syllable "and", so often used apathetically to string words together, refers here to an inextricable community.

*Karl Haushofer*, 1935

## In Defence of the Earth

In 1951, the former German Wehrmacht general Heinz Guderian published a short pamphlet in which he set out to debate Germany's destiny in the early Cold War. The book's title, which can loosely be translated as *Not Like This! A Contribution to the Debate of West Germany's Situation*, displayed a frustration with postwar politics that was typical for the German military elite. Germany had of course lost two wars within the span of a mere 30 years.

Guderian's book is immediately striking for its geopolitical objective, to write 'a political geography of our small terrestrial globe' (Guderian 1951a: 34).

> Let's begin with Central Europe, where we see our mutilated fatherland, bereft of its farming areas. The [territorial] rest of Germany is split in two, even the capital of the Reich is divided, its western sector a mere island in the Soviet zone. All German highland areas, the great rivers with the exception only of the Rhine, are in the Soviet zone (ibid.).

*Cryptic Concrete: A Subterranean Journey Into Cold War Germany*, First Edition. Ian Klinke.
© 2018 John Wiley & Sons Ltd. Published 2018 by John Wiley & Sons Ltd.

Of course, Guderian was no novice to questions of geopolitics. Not only had he been one of the pioneers of *Blitzkrieg* tactics, but he had also played a crucial role in the Third Reich's efforts to conquer living space itself, his troops failing to capture Moscow in 1941/42. Indeed, in 1937, he had written that

> We live in a world that is ringing with the clamour of weapons. Mankind is arming on all sides, and it will go ill with a state that is unable or unwilling to rely on its own strength. Some borders are strong, affording them complete or partial protection against hostile invasion, through chains of mountains or wide expanses of sea. By way of contrast the existence of other nations is inherently insecure. Their living space is small and in all likelihood ringed by borders that are predominantly open, and lie under constant threat from an accumulation of neighbours who combine an unstable temperament with armed superiority (Guderian 1937[1992]: 23).

Both before and after the war, Guderian expressed a geopolitical imagination that emphasised territory and the constraints of Germany's central position upon its foreign political options. Unlike many other West German observers, however, the former tank commander was comparatively critical of the Western Allies, accusing them both of cowardice vis-à-vis the Soviet Union and of attempts to punish Germany with a new version of the Versailles treaty. In clearly biopolitical tones, he warned that Germany would have to reunify its 'severed territories' if it wanted to be a 'viable entity' (*lebensfähiges Gebilde*). 'Too many wounds are on the body of our nation [*Volkskörper*] as that it could recover in its current state' (ibid.: 82–3). And yet, despite these accusations, his allegiance was fundamentally with Western civilisation and with the 'white race' (ibid.: 72). Although Guderian was uncertain as to whether the 'Eurasian or the western power group' would ultimately prevail in the Cold War, he held that Germany's centric position made a Soviet invasion likely, especially if the new government in Bonn were to pursue a policy of neutrality. This next war, he predicted, 'would be a "total war" in depth, width and height of the battlespace, drawing in all relations of human life' (ibid.: 25).

Striking about Guderian's pamphlet is not just the geo- and biopolitical nature of his arguments, but the extent to which he reframed the earth as a place of security, survival and salvation. Indeed, Guderian concluded that in the atomic age, 'the defence of the earth' had to be fought by building vital industries in a both decentralised and, crucially, subterranean way. '[N]o house, no industrial plant, no railway station and no public building shall be built without an air-raid shelter' (ibid.: 18–19). Whilst Guderian would play only a minor role in the eventual re-militarisation of West Germany (he died in 1954, a year before the Bundeswehr was founded), his writings during the early 1950s (see Guderian 1950 and 1951b) would serve to adapt the concepts of 1930s German geopolitics to the Cold War. As we will see in Chapter 3, Guderian was not alone in contributing to a renaissance of geopolitical ideas in West Germany. Interestingly,

Guderian's short book was published under a new series called 'Supplements in geopolitics'. The series was with the Kurt Vonwinckel publishing house, the same publisher that was also editing the freshly reborn *Zeitschrift für Geopolitik* (Journal for Geopolitics). It was in the pages of this journal that former members of the Nazi military elite would congregate in an attempt to make sense of the Cold War geopolitically. Indeed, former Wehrmacht officers such as Friedrich Ruge, soon to be the commander of the postwar German Navy, or Erich Hampe, the first president of the Federal Agency for Civil Defence, would publish work along similar geopolitical lines as Guderian. Crucially, this new West German geopolitics was transatlantic rather than transcontinental in orientation and concerned not with the conquest of territory as much as with the construction of *Lebensraum* in subterranea. Ultimately, this rebirth of German geopolitics would come to legitimate West Germany's nuclearisation and pave the way for its membership within the North Atlantic Treaty Organisation. In order to understand this remarkable transformation – and ultimately the re-emergence of the two spaces that materialise a geopolitics of survival and extinction, the bunker and the camp – we will have to travel back to the late nineteenth century and to the origins of geo- and biopolitics.

As was briefly discussed in Chapter 1, geopolitics first set out in the late nineteenth and early twentieth century to examine the causal influence that a state's climate, position and access to natural resources had on its rise and fall. Like other geopolitical traditions, German geopolitics was not designed as an abstract or detached ivory tower theory.[1] Germany's geopoliticians were enthusiastic nationalists, too, who saw their theory as both descriptive and prescriptive. They were designed 'to awaken Germany to the enormity of the stakes in the current struggle for colonial acquisitions' (Bassin 1987b: 480). Yet, despite being intended to be politically prescriptive, these theories of geopolitical space were ultimately too abstract to be turned into a political programme in any straightforward way. Theorists of *Lebensraum* had always been conceptually 'vague' about their understanding of the term (Murphy 1997: 245) and lacked a comprehension of how living space was to be administered *on the ground*. When the Third Reich eventually tried to turn geopolitics into policy, it did so by turning this politicisation of life and earth into architectural designs that would materialise these ideas on a micro-level.

This chapter begins by returning to the works of key geographers associated with German geopolitics in an attempt to show that we can find within their work the seed of thanatopolitics (the politics of death), understood here as the intellectual basis for a policy of national survival and extermination. I suggest that this simultaneously vital and morbid political geography already carried the seed of a spatial logic that would materialise in architectural form in the 1940s. For whilst the Third Reich tried to create a whole range of geopolitically themed buildings that attested to its imperial ambition, it is especially in the complementary architectural archetypes of the bunker and the camp that we can find in crystallised form an attempt to forge geo- and biopolitics out of concrete and iron. By looking in detail at the correspondence between spatial ideology and material politics, this chapter lays the foundations for an appreciation of the return of this spatio-material politics in the Cold War. For geo- and biopolitics did not simply disappear in 1945 but merely assumed a different guise. They went underground, re-emerging again in the complementary structures of the nuclear bunker and the atomic missile silo.

## The Rise of German Geopolitics

As we have learned in Chapter 1, the story of German geopolitics is usually told as a Faustian deal between a German geographer and the Third Reich, in which the former came too close to state power. The protagonist of this story is Karl Haushofer, a former major general in the German army turned academic. Whilst initially a regional expert on Japan, Haushofer and his prolific writings would increasingly operate on a more global scale, in the course of which he would emerge as the figurehead of the German geopolitical tradition. Much of Haushofer's motivation to write geopolitically was fed by a deep frustration with the European order in the aftermath of World War I. He rejected the terms of the Versailles settlement and desired for Germany to break out of its overpopulated *Mittellage* (centric position) and establish a greater pan-Germany. Developing further a number of existing geopolitical concepts such as *Lebensraum*, autarky and the land vs sea power binary, Haushofer would become centrally concerned also with the idea of pan-regions as well as with Germany's overpopulation and its 'unfavourable population density' (*ungünstige Siedlungsdichte*) (Haushofer 1926: 532).

Haushofer would perhaps have remained a marginal figure had he not in 1924 been introduced to an as yet comparatively unknown Austrian politician, Adolf Hitler. Haushofer met Hitler through one of the German geographer's students, Rudolf Hess, who would later become the deputy of the Nationalsozialistische Deutsche Arbeiterpartei (NSDAP) at the time when Hitler was putting together what was to become his autobiographical manifesto, *Mein Kampf*. Through Hess, Haushofer would later also meet with the German foreign minister Joachim von Ribbentrop, propaganda minister Joseph Goebbels and the Reich leader of the SS Heinrich Himmler (Heske 1987: 143). As a consequence of these meetings, Haushofer was credited in the United States as having had a formative influence on the writing of Hitler's book and thereby with inserting geopolitics into the heart of Nazi ideology.

Others, however, have held that Haushofer's influence was in fact rather marginal, especially after Hess flew to Britain in a failed attempt to broker a peace between the two Aryan races and was subsequently branded a traitor in the Third Reich. By 1941, Haushofer is commonly seen to have lost his influence on the Nazi elites (Herwig 1999: 236; see also Murphy 1997: 23).[2] One of the reasons for this was that Haushofer's theories displayed an ambivalence towards the Third Reich's biological theory of race (Bassin 1987a). As a result, German geopolitics has often been somewhat disassociated from the darker of the Third Reich's episodes, including its politics of extermination.

Karl Haushofer's personal and family life took a turn for the worse after he was dropped by the Nazis after the outbreak of World War II. In 1946, Haushofer would terminate his own life. Whilst he may have paid the price for the ideas that are today associated with the infamous tradition of German geopolitics, it was in fact another man, who had died almost three decades before Adolf Hitler came to power, who had first developed a theory on the spatial relationship between earth and life. Often regarded as the founding father of modern political geography, the German zoologist-turned-geographer Friedrich Ratzel made a name for himself by articulating a theory of state expansion in a scientific, or rather pseudoscientific, register, by making analogies to the animal and plant world. For Ratzel, states were organisms that could grow and shrink and that were rooted in the territorial 'soil' on which they stood. Writing in dialogue with the social Darwinist ideas of his time, Ratzel saw state behaviour as a relentless struggle for 'survival' or 'being' (*Kampf ums Dasein*).[3]

According to Ratzel, this battle for survival was in fact a struggle for '*Lebensraum*' – living space (Murphy 1997: 8). *Lebensraum* remains until this day perhaps the case of a 'disgraced concept', not just in Germany but elsewhere, too (Abrahamsson 2013: 37). Whilst the exact origins of the term are still the subject of debate, it is usually attributed to Ratzel (Halas 2014; Jureit 2012: 148). Comparing the movement of non-human life to that of nations, Ratzel saw the world as a vast array of touching and overlapping living spaces. '[E]very creature', he claimed, was 'spatially bound' (Ratzel 1901: 146) and thus his discussion of living space often featured long excursions into botany and zoology and included discussion of a wide array of species, including caterpillars and primroses. Ratzel saw this struggle for living space emerge out of the tension between the ever-expanding movement of life and the always stable and finite earth on which life dwelled (Ratzel 1901: 153). Combining Darwinian ideas and a Malthusian concern with overpopulation, he emphasised the role of vast and expansive livings spaces. Given that he saw the European continent as overpopulated, he stressed Germany's need to find new living space. Although writing in a period of rapid industrialisation, Ratzel's fantasy of *Lebensraum* was at heart an agricultural one (Smith 1980). The true potential of the German nation, he believed, was to be unlocked only 'through a policy of establishing and settling colonies, where true German culture could be recreated and preserved in an agricultural setting no longer attainable in rapidly industrialising Germany' (Bach & Peters 2002: 4).

Whilst Ratzel's ideas had found only modest support during his lifetime, they rose to prominence in the first half of the twentieth century. Amongst his followers were not just Karl Haushofer but also the Swedish political scientist Rudolf Kjellén (1864–1922) who, given the proliferation of his work in Germany, is usually considered a key figure in the tradition of German geopolitical thought (Tunander 2001: 451). Ratzel's concept of *Lebensraum* would help Kjellén to formulate his own organic state theory, the essence of which was that the political was 'nothing else but the continuation of nature at another level and therefore destined to incorporate and reproduce nature's original characteristics' (Esposito 2008: 17). As Abrahamsson (2013: 40) has observed, it was Kjellén's operation- alisation of Ratzel's ideas that established the conditions under which Haushofer would later popularise those same ideas.

Both Kjellén and Haushofer emphasised the role that agriculture would have to play in a nation's struggle for self-sufficiency (Herwig 1999: 226). Rather than relying on trade with others, the state should encourage a national economy that was so domestically intertwined as to produce a state of self-sufficiency. For Kjellén, the nation-state expressed the harmony between land and people, a space that would allow economic autarky. In his book *Der Staat als Lebensform* Kjellén argued that a nation should 'if necessary' be able to survive autonomously – 'behind closed doors' (Kjellén 1917: 162). Haushofer argued in 1934 that 'almost 3,000 years of history had crystallised the experience that world political independence und resilience are only to be sustained [...] if the basis of self- sufficiency is preserved' (Haushofer 1934a: 170).

The tradition of German geopolitics culminated in the thought of the philos- opher and legal scholar Carl Schmitt (1888–1985) who turned in the late 1930s to the question of geopolitics, even though he avoided that term. Unlike Haushofer, who often displayed a rather crude environmental determinism, Schmitt was perhaps a more considered spatial thinker whose understanding of political space some say foreshadowed the late-twentieth century spatial turn in critical social theory (Minca & Rowan 2015a). Though not technically a theorist of *Lebensraum*, Schmitt bequeathed to the world geopolitical concepts such as his reformulation of *Grossraum* (greater space) and *Nomos* (spatial order) (Minca & Rowan 2015b: 277). Grappling with the relationship between space and order, Schmitt was keen to underline that spatial revolutions, such as the crusades, touched not just the spatial measurements and horizons of societies but their very conceptions of space (Schmitt 1942[2001]: 55). There was, Schmitt argued, a relationship between the 'manifold forms of life' and variations in spatial con- sciousness that could be found between individuals and nations. Similarly to Haushofer's ambivalence to the Nazis' biological racism, Carl Schmitt's concept of greater space did not depend on the Third Reich's blood and soil racism (Elden 2010: 19; see also Minca & Rowan 2015b: 277). There was, in other words, a tension between space and race at the heart of the geopolitical project that ham- pered its influence (Bassin 1987a).[4]

*Lebensraum* resonated well with the political climate of post-World War I Germany where territorial revisionism was popular and nineteenth-century imperial fantasies had all but disappeared (Heske 1986: 270). Soon, the term had found its way into National Socialist doctrine. Echoing both Ratzel and Haushofer,

Hitler wrote in the pages of *Mein Kampf* that only 'a sufficiently large space on this earth' was able to 'assure the independent existence of a people' (Hitler 1925[1939]: 491). His posthumously published *Zweites Buch* ('Second book') stated in an unmistakably Ratzelian tone that

> The types of creatures on the earth are countless, and on an individual level their self-preservation instinct as well as the longing for procreation is always unlimited; however, the space in which this entire life process plays itself out is limited. It is the surface area of a precisely measured sphere on which billions and billions of individual beings struggle for life and succession. In the limitation of this living space lies the compulsion for the struggle for survival, and the struggle for survival, in turn, contains the precondition for evolution (Hitler 1928[2006]: 8).[5]

Given his assessment that the future of the German population depended on expanding its living space, Hitler proposed what he called a 'clear, farsighted policy of space', which secured 'a way of life for [the German] people through the allocation of adequate *Lebensraum* for the next one hundred years' (Hitler 1928[2006]: 159). As this vital space could 'lie only in the East', Germany's naval aspirations would take a 'back seat' whilst it attempted 'to fight for its interests by forming a decisive power on land' (ibid.). Hitler shared Haushofer's preoccupation with agriculture and held that there were 'vast spaces' on earth that were as yet uncultivated and that were 'only waiting for the ploughshare' (Hitler 1925[1939]: 115). 'Such land', he wrote, was there for a people with 'the strength to acquire it and the diligence to cultivate it' (ibid.).

Despite its reputation as the intellectual inspiration behind Nazi expansion, the influence of German geopolitics on the Nazi leadership may have in fact been rather limited. As we have already noted, its ideological standing was compromised in particular by an unwillingness to take on board the Third Reich's biological racism. Whilst the texts of Ratzel and Haushofer are thus commonly deemed to have prefigured the politics of imperial conquest, they are not seen as having foreshadowed the project of Nazi biopolitics. In this way, Haushofer's work is often read by contemporary geographers as having been 'recruit[ed] into a deadly regime' (Barnes & Abrahamsson 2015: 66) but not as already having in it the deadly seeds of biopolitics. If we follow this line of reasoning a little too hastily we get the impression that German geopolitics was fundamentally divorced from the most destructive facets of the Third Reich. We may legitimately wonder whether its placing of *space over race* also meant that German geopolitics was indeed disconnected from Nazi biopolitics, the project of exposing some forms of life to death in order to protect the population/nation as a whole (Giaccaria & Minca 2016). There are a number of reasons that would suggest otherwise, not least that the two terms geo- and biopolitics had originated in the work of the same man, Rudolf Kjellén. Moreover, it is in fact the organic theory of the state that can help us understand how it was possible for Jews and other groups to be constructed as 'cancerous cells' within the body politic. Indeed, Werner Daitz, an NSDAP economist, would publish a book in 1943 entitled *Living Space and a Just World Order: Foundations of an Anti-Atlantic Charta* in an attempt to formulate a theory of *Lebensraum* that was in line with biological racism. The idea that every Volk was meant to have its own *Lebensraum* was of course interpreted to exclude the Jews (Daitz 1943: 22).

Finally, and most importantly, the traditional formulation of German geopolitics by Ratzel and Haushofer had never been concerned simply with the birth and growth of nations, but also with their ageing and death. Indeed, death, extinction and extermination were important leitmotivs within German geopolitical thought. The biopolitical, in other words, was always already present in geopolitical discourse. As Murphy reminds us, geopolitical writers of the interwar period had often described 'a decaying German racial-genetic pool locked in mortal biological combat with hordes of fertile Slavs who were eternally threatening the German *Lebensraum* from its vulnerable Eastern borders' (Murphy 1997: 36). It is these leitmotifs of death and extinction to which we shall now turn, for they reappeared – in distorted and inverted ways – in the early Cold War.

## Life and Death in the German Geopolitical Tradition

Whilst we can find in German geopolitics neither an appeal to inflict a genocidal war on Eastern Europe nor a call to engineer a programme of racial extermination, it is important to recognise that German geopolitics was not detached from the biopolitical forces that would become so central to the history of the twentieth century. The colonial fantasies entertained by the likes of Friedrich Ratzel were articulated at a time when the second German empire was actively seeking colonial acquisitions in South-West Africa. Openly supportive of Germany's attempts to seek out new *Lebensraum* in Africa, Ratzel published his *Lebensraum* essay only a few years before Germany's genocide of the Herero and Namaqua (1904–1907) in modern-day Namibia, a massacre that was arguably fought in the name of a struggle for space (Danielsson 2009). Moreover, even though Ratzel's successors never adopted the biological racism prevalent in Nazi ideology (Bassin, 1987a), they were centrally concerned with the question of biopolitics. When Kjellén first used the concept of biopolitics in 1920, he wanted it to capture both the physical and cultural dimensions of life (Kjellén 1920: 94). Geopolitics, for him, was the 'doctrine of the state as a geographical organism or as a phenomenon in space' (Kjellén 1917: 46). Haushofer rarely defined the term geopolitics (Heske 1987: 137), but when he did, his definition, too, was unmistakably biopolitical. Geopolitics, he wrote, was 'the science of political life forms in their natural living spaces', a science that tried to understand how these life forms were conditioned both by the earth and by historical processes (Haushofer 1925: 87). Elsewhere he warned against the fallacy of treating a state's border in merely legalistic ways and stressed its organic nature. To do so would be to miscomprehend its nature as a 'living organ, a battle zone like an individual's skin' (Haushofer 1934a: 160). Few of those who used the catchphrase *blood and soil*, he argued, were aware that the two were inseparably linked (Haushofer 1935: 11).

Similarly, and this is something that seems to have gone widely unnoticed in the academic literature on German geopolitics, death was indeed an important leitmotiv for the likes of Ratzel and Haushofer. Echoing through their writing was an intellectual absorption with the idea of extinction as well as an aesthetication of – even perhaps a subliminal desire for – death itself (thantophilia). As a discourse on survival, geopolitics was necessarily also a discourse on extinction because it included within it the possibility of *not* surviving. 'Creation and progress' in Ratzel's words assume 'decline and demise' (Ratzel 1901: 161). Indeed, he argued that 'nations die when they fail to assimilate' (Ratzel 1941: 97; see also Ratzel, 1897: 174). Kjellén's work, too, was concerned with death. Whilst he admitted that the very thought that his fatherland would one day cease to exist filled him with horror, he was nevertheless keen to argue that *all* states were eventually doomed to disappear. The trick, he implied, was to see this decay of states through a detached historical perspective so as to disperse any emotions of gloom. 'On the graveyard of world history', Kjellén wrote, 'the tombstones of fallen states remind us that states must go the same way that all men do' (Kjellén 1917: 204). In this, one could argue, German geopolitics was somewhat deviating from other nationalisms that tend to see the disappearance of their own nation as the 'ultimate tragedy' (Billig 1995: 4). Kjellén's key example to illustrate this process was the case of the Polish partitions in the late eighteenth century, which he described as 'executions' that one should not feel sorry for. 'Death', he explained, 'had nested at the very heart' of the Polish state in ways that had long destroyed 'any feeling of national belonging' (Kjellén 1917: 215). 'Our natural pity with [this state's] great suffering should not seduce us into missing the organic nature of the Polish fate', he argued, going on to explain that the death of a nation resembled 'a fleshly death' and was therefore 'eternal' (ibid.). As Billig puts it, '[t]the aura of nationhood always operates within contexts of power' (Billig 1995: 4).

In a similar vein, Haushofer argued that geopolitics was not for the fainthearted – i.e. those who could not look the deadly struggle between states for space in the eye. 'Life itself', he explained, 'is at stake in this game' (Haushofer 1934a: 261). Those 'who cannot live as if every day was their last' were following a lost cause (ibid.). If the world did not adhere to the principles of geopolitics, he predicted, it would 'raise life annihilating storms' that would 'bury megacities and nations alike' (Haushofer 1935: 43). *Lebensraum* was of crucial importance if a nation was to be spared such storms. 'The loss of soil and territory leads to the state and nation's downfall and their eventual annihilation; no state is conceivable without sufficient Lebensraum', Haushofer held (Haushofer 1934b: 558). In relation to Germany, he warned that the German Reich could ill afford to lose any more living space if its nation was not to 'bleed to death' (ibid.: 589). Indeed, Kjellén thought of space as so vital that he was willing to suggest that it was less problematic for a state to be subjected to a decrease in population than to lose territory (Kjellén 1917: 57).

Carl Schmitt, too, spoke with significant detachment of the annihilation of states and populations. Writing at the height of World War II, he lamented that his emotional contemporaries would 'only see destruction and death' in the current turmoil and suggested instead that we should appreciate the war simply as 'the end of the previous relationship of land and sea' (Schmitt 1942[2001]: 107). The violence of history was thus reduced to the eternal struggle between binary oppositions, in this case of land and sea. For Schmitt, who shared both a fascination with questions of sovereignty and the politics of earth (Schmitt 1922[2005]; 1950[1997]), the notion of death may not have been as central as it was for Ratzel. And yet, in his 1932 *Concept of the Political*, in which he outlined his famous friend–enemy distinction, he argued that 'only a weak people will disappear', leaving intact that timeless distinction of friend and enemy that emerges from the omnipresent possibility of war (Schmitt 1932[2007]: 53). In an attempt to describe and understand such a total opposition between friend and enemy, he argued for a conception of just war that recognised the state's 'right to demand from its own members the readiness to die and unhesitatingly kill enemies' (ibid.: 46). A life, he felt, that had 'only death as its antithesis [was] no longer life but powerlessness and helplessness' (ibid.: 95). As Kearns (2011: 84–5) puts it, for Schmitt 'a really living person must be struggling against others while to survive, a people must struggle against other peoples'. Schmitt was convinced 'that the new times required a total state and by this he meant a society where the friend–enemy distinction colonised all spheres of society' (Kearns 2011: 86). Schmitt, in other words, was naturalising the idea of an enemy that can and should be annihilated, an idea that stands both at the centre of the Nazi fantasy of *Lebensraum* and at the nightmare vision of a nuclear war.

It is perhaps no surprise that German geopolitics had developed such a rich tapestry of terms in the interwar period to denote the process of decay and dying, many of which would return in the geopolitical discourses of the early Cold War. Haushofer frequently spoke of the dissolution (*Auflösung*), evaporation (*Verdampfung*), melting (*Einschmelzung*), mutilation (*Verstümmelung*) and demolition (*Zerstörung*) of states and nations. Both Ratzel and Haushofer used the term *Vernichtung*, which translates into English as extermination, annihilation or total destruction. Ratzel described the colonisation of the Americas as a 'struggle for extermination' (*Vernichtungskampf*). 'The losers', Ratzel explained, were the native Americans 'who had only a weak foothold on the soil' (Ratzel 1901: 158).[6] By naturalising extermination as a process in world politics, German geopolitics – along with a whole range of social Darwinist theories – laid the basis for a policy that saw extermination as a strategy to ensure the nation's survival in a competitive environment. As Haushofer's contemporary Hennig put it, 'very old states sometimes had to be incapacitated' when they became 'unfit for life [*lebensuntüchtig*]' (Hennig 1935: 90). This often bombastic language of survival and extermination would return to the fore in the early Cold War, when it was adapted to the power struggle between the Soviet Union and the United States and the nuclear arms race.

Given this fascination with the politics of extinction, it is perhaps not unsurprising that geopolitics was preoccupied with deathly objects – such as ruins (Hell 2009; Hell & Steinmetz 2014). Of course, the ruin was an object of wider fascination in the nineteenth- and twentieth-century mind, and arguably remains so today, but it is striking how prevalent the metaphor was in German geopolitics (Haushofer 1934a: 96; 1944; Ratzel 1941). Ratzel in particular rated polities that had produced ruins civilisationally higher than those that had not. Already in his travels to North America during the 1870s he had been fascinated by the ruins of the civil war and of human settlements that had been built and abandoned alongside the new railways. Indeed, this experience of witnessing decay along the vital boundary of the American frontier had prompted him to conclude that life was 'not stronger than death' (Ratzel, 1876[1988]: 295). Later, he would write of 'ruin lands', where 'a whole culture had been annihilated without any new life blossoming from the ruins' (Ratzel 1941: 100). These lands of decline were always to be found, he argued, on the battlefields that emerged between greater natural and historical oppositions, such as steppe and fertile land, nomads and civilised nations, or Islam and Christianity. Ratzel was not able to conceal his admiration for these traces of death and destruction when he concluded that 'a life that leaves ruins is more valuable than one that disappears without a trace' (ibid.: 99). 'A nation that leaves humanity with witnesses of its existence [*Zeugen seines Daseins*], lives on in its works. All others are just dead, even if an ancient inscription conveys their names to us' (ibid.). This concern with ruins would play an important role both in the political aesthetics of the Third Reich and in the planning for nuclear war.

## From Abstraction to Materialisation

Despite the unwillingness of the German geopolitical tradition to take on a biological theory of racism, Ratzel and Haushofer's spatial obsessions did eventually manage to find their way into Nazi ideology. But in order to do so, they had to undergo a process of translation and adaptation, for German geopolitics was largely devoid of a concrete agenda that would turn the concept of *Lebensraum* into policy (Werber 2014: 43). Ironically, geopolitics – as an ostensibly materialist and spatial theory – lacked an understanding of the very materiality that subjected humans to space. In its preoccupation with natural materiality (mountain ranges, oceans etc.), geopolitics missed the role of cultural materiality; it neglected the architectural blueprint in favour of the geopolitical map.

Ratzel, for instance, spoke about architecture only in passing, but when he did, as in his discussion of city walls and ruins, he was interested in the disappearance of particular architectural forms rather than the ways in which architecture might be able to serve his political geography (Ratzel 1906: 441).[7] Haushofer depicted colonial buildings in his books but failed to discuss the role of architecture in *making geopolitics work on the ground* (Haushofer 1934a: 57).[8] Schmitt was the only one to think through the relationship between *Ordnung und Ortung* (order and positioning) more systematically and the way it extended into the realm of the built environment (Schmitt 1941(1991): 81). He highlighted that the German word for peasant (*Bauer*) had etymological links not to the act of farming but to the term for the building (*Bau*) in which the peasant lived – but even Schmitt was ultimately interested in much larger scales than those of the settlement or the individual building.

Whilst the built environment had thus only played a very insignificant role in the geopolitical imagination of Haushofer and Ratzel, the National Socialists recognised that *Lebensraum* had not merely to be conquered but to be *built* – with stone, concrete, steel and barbed wire. Geographical theorists like Haushofer would give way to more practical geographers, such as Walter Christaller, who were willing to get their hands dirty in the Nazi colonisation of *Lebensraum* (Barnes & Minca 2013). The carving of empire and geopolitics through the built environment was of course not an exclusive preoccupation of the Nazis (Atkinson & Cosgrove 1998; Driver & Gilbert 1998; see also Ó Tuathail 1994). And yet, from the reorganisation of existing cities and the planning of model settlements to the constructions of individual monuments and mausoleums, the reorganisation of space within National Socialism took on 'its own fetishistic logic and power' (Barnes & Minca 2013: 681).

> Questions of living space, of empty or overpopulated space, of measured space and its 'content,' of the philosophy and operation of space, and of the imagination of space [...] kept busy an army of Nazi academics, experts, technicians, opinion makers, politicians, and military personnel (ibid.).

It is thus no surprise that the Third Reich constructed a whole range of architectural spaces, many of which were designed with very specific ideas in mind; as 'words in stone' (Taylor 1974). There were those, such as the gigantic Nazi party rally grounds in Nuremberg, which were designed to host embodied performances of the 'reborn' nation (Hagen & Ostergren 2006). New 'Germanic' settlements of 20,000 inhabitants were to mushroom across the new *Lebensraum*. Others, such as Berlin's monumental Tempelhof airfield or the famous rocket-launching site in Peenemünde, where Wernher von Braun developed his V2, were built to underpin Germany's geopolitical aspirations in the air. Others again, such as the monumental Prora holiday resort on the island of Rügen, aimed to recharge the German population's energy for the bellicose tasks ahead.[9] The Third Reich's famous *Autobahns*, too, were in a sense a form of geopolitical architecture. Whilst they had been planned already before the Nazis seized power, the motorways became a crucial ideological infrastructure during the 1930s. Using poured concrete as an architectural medium, they 'made possible the realization of the German landscape, and of the German Volk, changing each in the process into something new' (Forty 2012: 68). The network of motorways was a way of 'reconnecting' the German people, much of which was urbanised, 'with the land, the soil' (ibid.: 63).[10] It was thus in the 1930s that the Nazis started to build roads along the Third Reich's territorial borders that had no function other than to bring the *Volk* nearer the experience of these borders (Haney, forthcoming).

The attempt to link the geopolitical map and the architectural blueprint was perhaps nowhere clearer than in the close relationship of the architect Albert Speer and the failed architect Adolf Hitler, both of whom were captivated by the ways in which social relations could be manipulated through space. Like the geopoliticians, Hitler, too, was concerned with the question of extinction and ruins. He bemoaned his compatriots' lack of concern for the ways in which 'so many peoples and states ha[d] already perished and even disappeared from the earth' (Hitler 1928[2006]: 38). 'Vast cities', he held, had 'fallen into ruins, with hardly enough rubble remaining even to indicate their location to current generations' (ibid.). The pages of *Mein Kampf* were filled with an anxious fantasy that 'future generations might gaze upon' Germany and see 'the ruins of some Jewish department stores or joint-stock hotels and think that these were the characteristic expressions of [German] culture' (Hitler 1925[1939]: 208). Germany, in other words, was seen by the Nazis to be lacking the kind of spectacular buildings that would tell future generations the story of the nation's magnificence.

Determined to construct buildings that would leave a lasting impression on future generations, Hitler saw in Albert Speer a man who would help him realise his geopolitically themed architecture. In collaboration with this man, whom he would in 1942 promote to Minister of Armaments and War Production, Hitler worked on designs that were meant to express the most fundamental ideas of National Socialism: its nationalism, its militarism and of course its geopolitical ambition. Speer would later remark in his memoires how Hitler had once shared his own sketchbook from the early 1920s with him and that he had noted that architectural sketches of public buildings 'often shared the page with sketches of weapons and warships' (Speer 1970: 78). Like the Nazi fantasy of *Lebensraum* itself (a peasant empire conquered by the most modern of military technologies), many of Speer's buildings were characterised by a tension between classicism and modernity, mysticism and rationality. They too expressed a cyclical understanding of history in which empires were seen to perpetually rise and fall. Most importantly, and again much like the fantasy of conquering *Lebensraum* in Eastern Europe, Speer's buildings lacked a sense of scale. The need for megalomaniac proportions was often expressed by Hitler in reference to the kind of structures seen in other Western European capitals and the United States (Thies 2012: 78).

They spoke, in other words, to the same national inferiority complex that had once haunted Karl Haushofer's writings on the post-World War I settlement and the disillusionment with Germany having been stripped off its great power status. This anxiety about Germany's geopolitical fragility would return, as we will see, in the West German discourses of the early Cold War.

Much has been written about Adolf Hitler and Albert Speer's plans for a *Weltstadt Germania* ('World city Germania'), an architectural fantasy for a rede-signed Berlin that would replace a city deemed too bourgeois, cosmopolitan and ultimately Jewish by the Nazis to be the nerve centre of the new European geopolitical order. Many of Speer's planned buildings, such as the gigantic 'People's Hall', made stylistic references to Rome and other ancient civilisa-tions. Their size invoked a world order in which the Reich's thirst for fast expan-sion had finally been quenched. Léon Krier has described this structure as 'motherly' and 'telluric' in its roundedness (Krier 1985[2013]: xx). Unlike many Soviet memorials or skyscrapers in the liberal capitalist regimes, it did not point to the sky but at the earth. Redesigned along a new boulevard (the North–South axis), Berlin was re-imagined as a 'continental metropolis', the capital of a land power whose orbit would 'stretch from the Atlantic Ocean to the Ural Mountains' (ibid.).

Like Hitler, Ratzel and Haushofer, Albert Speer was fascinated by ruins. Hitler's architect would later state that he had designed buildings that would ruin beautifully in centuries to come and retroactively termed this his 'theory of ruin value' (Speer 1970: 97). This architectural approach imagined a gaze from the future onto the present – that would make future generations worship Hitler's thousand-year Reich in the same way that the National Socialists romanticised ancient Rome.

> It was hard to imagine that rusting heaps of rubble would communicate the[] heroic inspirations that Hitler admired in the monuments of the past. My 'theory' was intended to deal with this dilemma. By using special materials and applying certain principles of statics, we should be able to build structures which even in a state of decay, after hundreds or (such were our reckonings) thousands of years would more or less resemble Roman models' (ibid.).[11]

As Featherstone (2005: 302) has noted, Speer's theory of ruin value was than-atophile 'in that it sought to trade the temporality of life for the eternity of death'. In this it was very much united with the deathly obsessions of Friedrich Ratzel. Indeed, the Nazi architect Wilhelm Kreis planned a string of necropolises that would be erected on the outer margins of the new *Lebensraum* (Michaud 1993: 227). Rather than simply designed to commemorate the dead warriors who had perished in the struggle for living space, these sites were to materialise Germany's geopolitical ambition and the idea that even if one day Germany was no more, its stones would speak to future generations.

We shall return to this ruin aesthetic in subsequent chapters, because it lives on in a different form in the fantasy of surviving a war that had left the world in radioactively contaminated ruins. But before we do that, we have to recognise that the Third Reich's attempts to construct representative architecture that would cast in stone its geopolitical and historical ambitions was actually a rather different architecture that would express the fantasy of *Lebensraum* and the politics of survival and extinction much more directly. Two complementary architectural archetypes are crucially tied up with the struggle for *Lebensraum*: the bunker and the camp.

## Complementary Archetypes

Perhaps the most crucial architectural archetype constructed by the Nazis in the quest for living space was in fact its exact opposite – the extermination camp, a space of industrialised killing. Concentration camps had been set up by the Nazis soon after their ascent to power in 1933, particularly for political enemies, but it was only after 1942 that the Third Reich moved towards the systematic extermination of large parts of the European population in camps. Of course, concentration camps were not invented by the Nazis but had already been used by the British in the Boer War and were developed further by the Soviet Union in its archipelago of slave labour camps. As Diken and Laustsen (2004: 17) remind us, camps are thus 'part and parcel of European history and cultural identity'. And yet, the Nazi death camps are to this day commonly remembered as the most perfidious camps in modern history. In camps like Belzec, Treblinka, Sobibor and Auschwitz, humans, the vast majority of which were Jewish, were rendered 'passive recipients of violence' in the sense that they were 'reduced to flesh' in becoming 'a mere biological receptacle for pain and disease' (Netz 2004: 130, see Figure 2.1).

Much of this happened through a very particular matrix of spatial power. Extermination camps were biopolitical ordering devices that worked to separate out and devalue some human life to the degree that it could be exterminated without repercussions for the exterminators. It is important to understand that the appearance of the camp within a state that was guided by the fantasy of conquering living space was no coincidence. The geopolitical fantasy of a vast and open living space necessitated the closed and concentrated space of the camp as a counter-dimension in order to 'cleanse' the body politic. The catchwords *Lebensraum* and *Judenrein* ('clean of Jews'), in other words, represented 'the two sides of the same spatial exercise – the outward and inward pushes of the same drive' (Netz 2004: 195). Although it is important to stress that geopoliticians like Friedrich Ratzel or Karl Haushofer had never conceived of anything even remotely resembling a concentration camp, it was in the extermination camp that the dying of nations that they had described and naturalised in their writings could happen in an accelerated way. Indeed, it is perhaps no coincidence that the first camps emerged precisely at a time in the late nineteenth and early twentieth century when the question of space was beginning to be fetishised by geographers like Ratzel.

Taking his cue from Carl Schmitt's theory of sovereignty (1922[2005]), Giorgio Agamben has argued that the camp emerges when the state of exception, that simultaneously legal and extra-legal moment that defines sovereignty, takes on a permanent and material form. Rather than the city or the prison, Agamben argued, it was the camp, that 'pure, absolute, and impassable biopolitical space' that constituted 'the hidden paradigm of the political space of modernity' (Agamben 1998: 123). It was here that sovereign violence could be inflicted on human bodies that had their political subjectivity removed from them, having been reduced to 'bare life'. Agamben held that the Nazis, who had established their rule through a whole series of emergency decrees, killed by declaring some forms of life 'unworthy of living', thus legitimating extermination in a biopolitical rather than religious or legal register.

> The camp, in this theory of sovereign power, is the actual space where citizenship may be arbitrarily put into question, where people are translated into mere biopolitical bodies. The camp as a political technology is precisely how de-subjectivation is made operational and possibly taken to its extreme manifestations, to the point of producing in the Nazi concentration camps, biopolitical individuals that are half alive and half dead (Minca 2015: 79).

Whilst generally useful in linking the camp to the state of exception and extrapolating the camp's wider significance, Agamben's preoccupation with the topological political logics that enable the camp obstructs a better understanding of the concentration and extermination camp's topography, its architectural and human infrastructure (Bernstein 2004; see also Gregory 2007: 210). Enclosed by barbed wire and watchtowers, the camp would have a prohibitive zone next to the external fence, which prisoners 'could not avoid' getting near to – thus living in constant fear 'of crossing the barrier and being shot' (Netz 2004: 214). Exhibiting a geometry of power, the highly constricted space of the camp used barbed wire not just to secure the outer parameters but also to delineate areas for administration, accommodation and extermination (ibid.: 210). As Giaccaria and Minca (2011: 5) have argued about Auschwitz in particular, the camp's rigid separation from its outside 'was paralleled by an obsessive calculative management of the interior'. 'The internal spatialities of the camp – from the dormitories to the latrines', they note, 'were planned in detail in order to minimize the consumption and the use of space, and to maximize control and discipline' (ibid.). In this highly confined space, Netz argues, humans were gradually transformed into human cattle. They were reduced to their function as basic organisms, branded and either worked to death or simply exterminated. Crucially, this process of extermination was at all key stages accompanied and enforced by members of the medical profession, from the selection on the ramp leading to the camp to the process of gassing itself (Esposito 2008: 113).

It is important to see the extermination of the European Jews and other populations in a network of camps as an extension of the earlier biopolitical strategy of forced euthanasia of humans deemed to be mentally ill or physically disabled under the so-called T4 programme. As Esposito (2008: 136) notes, the T4 programme was backdated to the outbreak of World War II and therefore presents 'the most obvious sign of the thanatopolitical character of Nazi biopolitics as well as the biopolitical character of modern war'. It is this blurring of the line between war and peace, life and death, that is also captured by Agamben when he argues that:

> If there is a line in every modern state marking the point at which the decision on life becomes a decision on death, and biopolitics can turn into thanatopolitics, this line no longer appears today as a stable border dividing two clearly distinct zones (Agamben 1998: 122).

Even where the Nazis' racial agenda gave way to questions of economic exploitation, the thanatopolitical drive was still in place. Death through forced labour was an important facet of the project of racial genocide in the new *Lebensraum*. Indeed, there was an intimate connection between the network of camps that mushroomed throughout the Reich and the monumental building economy that we have mentioned above. As Speer and Hitler's megalomaniac fantasies relied on forced labour in their realisation, new concentration camps were often being set up near quarries (Thies 2012: 101). The architecture of Hitler's thousand-year empire was thus to be made out of the resources of the new *Lebensraum*. Slave labourers from Eastern Europe would assemble it with granite from Scandinavia and limestone from France.

The second architectural space that defined the project of Nazi *Lebensraum* even more than Hitler and Speer's monumental fantasies was the bunker. From the 1940s onwards, it could be found both at the outer margins of the new European order and increasingly the fascist city. Bunkers had already played a crucial role in World War I but it was only during Hitler's twelve-year empire that they emerged in a wide variety of shapes (Bennett 2011a). The majority of bunkers were military bunkers, built to secure soldiers, submarines or industrial production. The most colossal undertaking within the Nazi bunker-building economy was the Atlantic Wall, which was to mark the empire's outer boundary in the West. Built between 1942 and 1944 along the Atlantic and North Sea coast,

**Figure 2.1** Auschwitz-Birkenau (1977). Source: Auschwitz barbed wire fence. Lars K. Jensen, 2008. https://commons.wikimedia.org/wiki/File:Auschwitz_barbed_wire_fence.jpg. Licensed under CC 2.0.

**Figure 2.2** Civilian bunker, Germany (around 1943). Source: Das Bundesarchiv. Reproduced with permission.

these bunkers were predominantly defensive in design.[12] As Albert Speer recalled after the war, it was Hitler himself who had planned some of these installations 'down to the smallest details'. Indeed,

> He even designed the various types of bunkers and pillboxes, usually in the hours of the night. The designs were only sketches, but they were executed with precision. Never sparing in self-praise, he often remarked that his designs ideally met all the requirements of a frontline soldier (Speer 1970: 477).

As Speer clarified, these designs were adopted almost without revision by the general of the Corps of Engineers (ibid.). Interestingly, even the designs for The Supreme Commando of Armed Forces and Hermann Göring's Reichsmarschallamt included command bunkers, despite the fact that they would have been completed only after a successful military campaign and the establishment of a new geopolitical order under Nazi control (Krier 1985[2013]: 95).

When it became clear that Germany was likely to be hit by extensive Allied bombing campaigns, the Nazi state started the systematic construction of a second type of bunker, the civilian shelter, which represented an archetypically fascist space (Figure 2.2). It was only beneath metres of reinforced concrete that the Nazi nation building project was finally realised. Regardless of their societal rank, Germans were united underground in an apparent state of security – whilst forced labourers and concentration camp inmates had to endure the bombing campaigns outside the concrete survival shell of the bunker (Friedrichs 2008: 247). Hitler's own underground retreats, such as the *Führerbunker* near the Reichstag in Berlin, were designed to protect the dictator and his entourage from the bombs. Given Hitler's suicide in what his staff would later describe as a 'coffin' (Fest 2005: 106), the bunker has become a symbol of the self-destructive drive of the Nazi state, which I will return to in more detail below.

It is important to understand that the bunker was not just an inevitable outcome of the genocidal war that the Third Reich had inflicted on the world, but very much part of the Nazi architectural aesthetic itself. Unlike in other countries, many of the earlier civilian bunkers were built to resemble castles, so as to blend in with the neo-medieval fantasy space of the Nazi city. These bunkers were very much part of the Wagnerian *Gesamtkunstwerk* (total work of art) that Nazi Germany had wanted to construct all along. And yet, in many ways, they were more of an artistic statement than a functional necessity. Indeed, the bunkers on the Atlantic coast did not withstand the Allied landing. The civilian bunkers, too, were prone to malfunction. During the firestorms that were ignited by Allied bombing, many of the bunkers would be turned into furnaces, in which their inhabitants were quite literally baked to death.

It is in this context that we need to return to Paul Virilio's (1975) theorisation of the bunker as a spectacular yet insecure monolith that promises survival in an age in which weapons have become so omnipotent that distance can no longer act protectively. We have already learned in Chapter 1 that the bunker emerged through the technological possibility and the political will to take out whole cities. Like the camp, it can be read as a modern excess, 'the waste of modernity that cannot be tidied away' (Beck 2011: 83). Similar to the 'anxious urbanism' of Cold War era urban planning (Farish 2004: 94), nuclear bunkers materialised both the fear of nuclear war and a radical disillusionment with urban life. Like camps, they functioned as 'anti-cities' that denied the experience of cosmopolitan space (Virilio & Lothringer 1983: 132).

After 1945, bunkers were often amongst the very few structures in Germany that were still intact. Given their military significance, the bunkers were intended to be blown up by the Allies, despite the fact that they continued to play an important role in sheltering refugees and those who had lost their homes to the bombing campaigns (Friedrichs 2008: 246). The Allies were willing to make an exception only where bunkers could be deprived of their function, for instance by drilling windows into the concrete. Soon, the civilian Nazi bunker landscape seemed to have been dismantled.

## Autoimmunity

Once the Wehrmacht had embarked upon its expansion into the European East, the question of living space was no longer merely an academic or ideological one but one of governance, too. As the reorganisation of European space and resettlement of vast populations was starting to be realised, it soon became clear that German geopolitics was ultimately a meta-theory. Rather than providing a blueprint for the establishment of empire, it framed the intellectual parameters of political debate and directed the ideological struggle towards a very specific aim, the establishment of national survival through a policy of territorial expansion and racial extermination. Even Hitler's *Mein Kampf* was hardly detailed and practical enough to be 'the last word' and thus the establishment of *Lebensraum* had to be 'worked out in response to actual experiences of occupation' (Housden 2003: 106). Similarly, the European *Grossraum* had to be shielded both from the bombing campaign that was steadily intensifying and against the possibility of Allied invasion on the Atlantic coast.

As Allied bombing and the extermination of the European Jews entered its crucial stage after 1942, the two architectural archetypes of the bunker and the camp started to play a pivotal role. We are reminded here again of Bratton's insistence that both spaces attempt a radical inside/outside distinction, but whilst one functioned as 'an architectural membrane against a hostile world', the other would perform 'an expulsion-by-enclosure of the Other from the normal performance of law' (Bratton 2006: 19). In their entanglement with questions of survival and extinction, I have argued in this chapter, both spaces also express the political horizon of the German geopolitical tradition.

Both the bunker and the camp were architectural spaces constructed by a state that was so destructive that it had turned onto itself. From the late 1930s onwards, the Third Reich had witnessed a will to annihilate humans and materials at an unprecedented level. As it became clearer that Nazi Germany would be likely to lose the war, the genocidal campaign was increasingly directed against the state itself and its population. It was the realisation that the war was lost and the Atlantic Wall would be breached that prompted propaganda minister Joseph Goebbels to proclaim a total war that would realise ruin value in only a few – rather than a few thousand – years (Virilio 1975: 58). The so-called Nero command, issued in March 1945, stipulated that in this 'struggle for existence' in which the nation was involved all 'military traffic, communications, industrial and supply installations as well as objects within Reich territory that might be used by the enemy in the continuation of his fight, either now or later, [were] to be destroyed' (Hitler 1945).

These last orders are commonly interpreted not as defensive measures but as the revelation of the Third Reich's death wish. The policy was in effect an extension of the Wehrmacht's 'scorched earth' policy, which had already begun on the retreat from the Soviet Union. Hitler himself had told foreign visitors that he thought the German people should be 'annihilated' as they were 'no longer strong enough and willing to shed their own blood to ensure their survival' (as quoted in Fest 2005: 130). Hitler is also reported to have added that he would not 'shed a single tear for them' (ibid.). By destroying the German economy and military infrastructure the Third Reich was trying to prevent a national rebirth. It was as if the Nazis were intervening into and exacerbating the forces of national survival and decline that had been sketched by Ratzel only a few decades earlier.

As Roberto Esposito has argued in his re-reading of the question of bio- and thanatopolitics, modern life is immunised through the production of death. He aptly compares the Nazi state's suicidal impulse to an autoimmune disease that turns the protective apparatus against itself. The political space in which this happens most clearly for Esposito is the bunker:

> The final orders of self-destruction put forward by Hitler barricaded in his Berlin bunker offer overwhelming proof. From this point of view, one can say that the Nazi experience represents the culmination of biopolitics, at least in that qualified expression of being absolutely indistinct from its reversal into thanatopolitics (Esposito 2008: 10).

Esposito concludes that the Nazis proposed a solution to the presence of death in life. 'The only way for an individual or collective organism to save itself definitively from the risk of death', he explains, 'is to die.' This, of course, was exactly 'what Hitler asked the German people to do before he committed suicide' (Esposito 2008: 138). Thus, to quote Featherstone, the 'thirst for annihilation [...] turned upon the humanity of the Nazi self' (Featherstone 2005: 305). It was, in other words, 'only through the destruction, ruination, and death of the fascist self that entry into the imperative sphere of order, eternity, and certainty could be made secure' (ibid.).

It is important to acknowledge that the preoccupation with questions of survival and extinction does not necessarily point towards the holocaust in any straightforward way. Similarly, geopolitics was not alone in its preoccupation with the question of extinction. And yet, German geopolitics created an intellectual milieu in which it was possible to think about extinction in certain ways, in ways that would enable the Nazis to move from merely observing a struggle for national survival and extinction to an attempt to shape that very process. This becomes especially important as we turn to a period in which there was seemingly no intent to exterminate, in which the darker aspects of the Third Reich's rule over Europe had been cast as something that should never happen again. And yet, it is precisely in the aftermath of World War II and the rise of new and omnipotent weapons systems that we see an obsession with space, survival and extinction re-emerge, both at the level of political thought and at that of the architectural blueprint. The Nazi project of *Lebensraum* was not the only project in the mid-twentieth century that was obsessively concerned with the question of space. As we will see in what follows, the Cold War, too, was preoccupied with geopolitical mappings – and it too was committed to the construction of biopolitical spaces.

After 1945, both German geopolitics and political geography more generally were discredited. And yet, as I will argue in what follows, the logic of national survival and extinction that had been at the heart of German geopolitics managed to endure into the Cold War, not just by re-emerging in the discourse of nuclear strategy, but also by materialising in the self-destructive space of the tactical nuclear weapons camp. In developing this argument, I trace how an obsession with autarky and survival led to the construction of nuclear bunkers that in many ways reproduced and inverted the political space of the concentration camp. As we will see in subsequent chapters, the nuclear bunker, as a further development of the Nazi bunker, and the nuclear missile camp, as a further development of the death camp, turned the biopolitical logic of the camp inside out. Whereas the Nazi empire's Eastern living space was punctuated by concentration and death camps, the subterranean living spaces of the nuclear bunker would be scattered across a potential landscape of Cold War extermination. Whilst the crematoria and gas chambers were directed towards the inmates, the Cold War's tactical nuclear weapons were pointing at the population all around them.

## Endnotes

1   German geopolitics never stood alone in its endeavour to simultaneously explain and shape world politics. The British geographer Halford Mackinder conceived of his political geography as a practical form of knowledge that would extend its reach beyond the academic ivory tower and educate the masses (Kearns 2010).

2　As Mark Bassin has argued, the idea that Haushofer was the man behind Hitler was 'untenable' and the relationship between geopolitics and the Nazi state was 'uneasy' at most. 'For the Nazis', he has held, 'the individual-imbued with a specific set of racial characteristics was supreme, while Haushofer and his colleagues stressed environmental influences. For them, considerations of inherited genetic qualities were, if not entirely irrelevant, at least clearly not of primary importance' (Bassin 1987a: 116).

3　For more biographical accounts of Ratzel, see Wanklyn (1961) and Müller (1996).

4　Schmitt continues to be a subject of controversy in human geography as it is debated as to whether there is still something that contemporary geography can learn from his oeuvre (Elden 2010; Legg 2011; Minca & Rowan 2015a and 2015b).

5　It is important not to overstate the continuity between the geopolitical ideas developed by Haushofer and their reception by the Nazis. Whilst Haushofer had advocated a continental block with Russia and Japan, Hitler wanted to gather *Lebensraum* in the East, a policy that was incompatible with Haushofer's dream of a Russo-German alliance (Heske 1987: 136). Nevertheless, this should not obstruct the view of the ideational congruence between the academic and policy versions of *Lebensraum*.

6　It is important to note that unlike the term *Aussterben* ('extinction'), *Vernichtung* assumes an agent. As in English, extinction takes a passive construction but extermination is an active process.

7　One of the exceptions here is his book *Sketches of Urban and Cultural Life in North America*, originally published in German in 1876 (Ratzel 1876[1988]), which does discuss the built environment in US cities. The book is however devoid of an interest in geopolitics.

8　Territorial boundaries were sometimes discussed in reference to particular natural material features (such as mountain ranges or rivers) – but they were not discussed as social systems that had to be administered through assemblages of bodies and cultural materials.

9　Other facets of the Nazi spatial fantasy, such as the restoration of the medieval city centre in Nuremberg, were perhaps less spectacular. But here, too, geo- and chronopolitical goals were at play (Hagen & Ostergen 2006).

10　In this sense, concrete was appropriated as a German rural and *Völkisch* material that stood in contrast to other materials, such as steel, which was identified as urban and essentially American.

11　It is important to remember that Albert Speer's theory of ruin value was a retroactive theorisation rather than a part of National Socialist doctrine (Fuhrmeister & Mittig 2008).

12　There were, however, exceptions, such as the bunkered ballistic missile launch site known as La Coupole ('the dome') in northern France.

# Chapter Three
# Return to the Soil

*The Russian is a kind of nocturnal animal, resourceful and without nerves. [...]*
*He will disappear into the earth like a clever mole.*

Leo Gyr von Schweppenburg, 1952

## Jumping the Big Pond

As we have seen in Chapter 1, it has been common to assume that geopolitics disappeared in Germany with Karl Haushofer's suicide, reappearing only on the other side of the Atlantic and feeding into the preparation for Washington's emerging conflict with the Soviet Union. Indeed, in the young Federal Republic, the terminology of *Geopolitik* was placed under a ban. If the West German state was to be trusted by its Allies, it had to avoid being associated with the expansionist project of *Lebensraum*.

In the first decades after the war, it was thus common for politicians to come under attack for speaking too self-evidently in a Ratzelian or Haushoferian manner. In 1955, for instance, the former Nazi and future West German chancellor Kurt Georg Kiesinger was accused of perpetuating the logics of *Geopolitik* in the West German parliament. The Conservative member of the *Bundestag* had argued that the ongoing global population explosion and technological advances were making 'Western European spaces life-threatening' (Bundestag 1955a). In the same year, the Federal president Theodor Heuss made sure he used the adjective 'geopolitical' only in quotation marks, as his critics in the reborn journal *Zeitschrift*

*Cryptic Concrete: A Subterranean Journey Into Cold War Germany*, First Edition. Ian Klinke.
© 2018 John Wiley & Sons Ltd. Published 2018 by John Wiley & Sons Ltd.

*für Geopolitik* bemoaned (no author 1955). The opposition in parliament to successive Conservative-led governments was in particular against what it referred to as 'geopolitical determinism'. 'One always hears about our geographical position, which we cannot alter', the Social Democrat Carlo Schmid argued in 1958, 'but geography should be no more than a resource for politics' (Bundestag 1958).[1]

On the other side of the Atlantic, it was possible to be more unambiguously geopolitical. Already during the early 1940s US elites had become increasingly fascinated with the idea that German geopolitics had been the fundamental driving force behind Nazi expansion, that it had indeed been the energy that had fuelled a military campaign that brought most of the European continent under temporary German control. It was in this context that a colourful myth of a secret geopolitical institute had emerged both in the US print media and on screen (Ó Tuathail 1996; see also Murphy 2014). By exaggerating the influence of geopolitics on Nazi foreign policy, these accounts had created the desire for an American version of this powerful mode of thought.

In this vein, American writers like Robert Strausz-Hupé, a key voice in urging the United States to join the war against Hitler, had managed to communicate this desire for geopolitics (Strausz-Hupé 1942). Geopolitics, he felt, was something that the United States did not yet but very much had to understand if it wanted to play a more active role in the international arena. The consequence was a remarkable exercise of intellectual import. In the same way as Haushofer had learned from the British 'enemy' geographer Mackinder, so were the Americans learning from their wartime German enemy (Ó Tuathail 1996: 130). And yet, geopolitics was disseminated in the United States with a twist. Increasingly, Geographers like Isaiah Bowman promoted a liberal version of geopolitics. Given the increasing tendency to frame geopolitics as a Nazi science in the American press, Bowman, then an advisor to President Roosevelt, felt compelled to delineate his own version of geopolitics from what he perceived to be the pseudoscience of German geopolitics (Bowman 1942). In doing so, however, he still hung on to the idea of *Lebensraum*, arguing that there was 'virtue in the argument of "organic boundaries" and the philosophy of Lebensraum', even if they were 'open to abuse'. The key, Bowman noted, was to reduce *Lebensraum* to its economic meaning and strip it of its connections to territorial conquest (ibid.: 656; see also Smith 2004).

After 1945, geopolitics was pushed to the next level as a new anti-Soviet foreign policy emerged that framed the enemy in an unmistakably geopolitical register. A key figure in this transition from World War II to the Cold War was the diplomat George F. Kennan whose anonymous 'Mr X' article in *Foreign Affairs* asserted an interpretation of the Soviet Union as an inherently despotic, aggressive and expansive force (Kennan 1947).[2] Carl Schmitt, too, made a clandestine revival in US Cold War geopolitics, particularly in the work of his disciple Hans Morgenthau (Coleman 2011). Although the realist Morgenthau thought of geopolitics as a pseudoscience (Morgenthau 1948[1993]: 174), he believed, much like the German geopoliticians, that states had an inherent desire for power and that the national interest was always to be privileged over more universal considerations.[3] Keen to adopt Schmitt's critique of liberalism and his understanding of sovereignty, Morgenthau also took on board Schmitt's concept of the political, which, as we have seen, was based on the possibility of a fundamental enmity, of a struggle between life and death (Williams 2005: 86). Perhaps most interestingly, Morgenthau was also centrally concerned with the idea of 'national survival', which he would famously elevate to a 'moral principle' (Morgenthau 1948[1993]: 12). Much like the German geopoliticians, Morgenthau remained preoccupied with the idea of extermination (ibid.: 228), even though he did not look at it in the same detached manner as had the German geopoliticians. A consultant to the US Department of State under Kennan in the 1960s, Morgenthau was later to be read by generations of students and was thus to have a pervasive influence on US elites (Kuklick 2006: 72).

Perhaps the most important figure to promote geopolitical ideas in the context of the Cold War was the academic-turned-secretary of state Henry Kissinger. Indeed, it was Kissinger who popularised the term geopolitics to the degree that it increasingly just stood in for 'global politics', 'power politics' or the territorial dimension in international politics (Ó Tuathail 1996: 17). Kissinger's 1957 book *Nuclear Weapons and Foreign Policy* is of particular importance in the evolution of American nuclear strategy. Read on both sides of the Atlantic, the text was critical of the all-or-nothing nuclear strategy of the Eisenhower administration, making a case instead for a selective use of smaller battlefield nukes that would enable the two superpowers to enter into a limited rather than apocalyptic all-out nuclear war. The United States, he argued, had to see the 'opportunities' and not just the risks of a nuclear war (Kissinger 1957[1969]: 13). This conception of a limited nuclear war would become crucial for West Germany, the designated battlefield of this war, as we will see in what follows.

The journey of geopolitical ideas from Germany to the United States is now a familiar story, and one that is commonly told as a tale of survival and loss. Geopolitics managed to cross the Atlantic by losing something, namely biopolitics. North American Cold War geopolitics, in other words, is often deemed to lack the biopolitical underside of a Ratzelian and Haushoferian geopolitics (Werber 2014: 143). It is important to note that Friedrich Ratzel, who had once been endorsed by the first female president of the Association of American Geographers, Ellen Semple (Semple 1911), did not go entirely out of fashion in the United States after World War II. Not only were Ratzel's agricultural theories popular amongst those trying to understand the nature of the Soviet state (Dalby 1990b: 76), but geopoliticians like Saul B. Cohen continued to cite Ratzel authoritatively until at least the 1970s (Cohen 1975). But whilst Darwinian conceptions of natural selection indeed lived on in the thought of Cold War American neorealists such as Kenneth Waltz (1979), it is true that American Cold War geopolitics did not depend on an organistic understanding of the state.

In what follows, we will venture beyond the now familiar story of how geopolitics experienced a renaissance in US foreign political discourse of the early Cold War period and instead observe its resurgence in the Federal Republic during the 1950s and 1960s – paradoxically despite being tabooed at the crucial level of official political discourse. We will discover that West German military strategists and civil defence planners tried to adapt the concepts of interwar German geopolitics to the Cold War. In doing so, they departed from the assumption that the Soviet Union and the Red Army had proved victorious in the previous war for two reasons, a willingness to look death in the eye and an ability to endure even in the harshest of climates. Made up of members of the former Wehrmacht, the new German geopoliticians were similarly concerned with Germany's unfavourable geographical position and territorial status after the 1945 Potsdam Conference, which had left Germany split in two and deprived of its Eastern territories (now part of Poland and the Soviet Union). Much like the Third Reich's spatial planners, these intellectuals of statecraft were increasingly concerned with political space not merely as cartographic or abstract space but as *architectural* space. Writing at length about the distribution of global power, these military strategists and civil defence planners were well aware of the limits of German power. In the first years after the war, the vision of conquering and building territorial living space in Europe's East thus gave way to the more modest fantasy of finding it in subterranea. This intellectual move entailed nothing less than a new conception of the earth as a place of salvation.

## The Rebirth of German Geopolitics

After the foundation of the Federal Republic in 1949, the young state's political elites had to be careful not to give the impression of seeking continuity with Nazi foreign policy. In the German Democratic Republic, too, Marxist–Leninist critiques of German geopolitics would soon emerge, in which geopolitics was read as a specifically imperialist obsession developed by the elites of an imperialist state (Heyden 1958). Although Chancellor Konrad Adenauer and his West German administration would stay clear of the most discredited geopolitical terms, this did not mean that geopolitical writing was censored *per se*. Indeed, the Vowinckel publishing house started to release a number of geopolitical works alongside revisionist books on World War II from the early 1950s onwards. In 1951, it also started republishing the *Zeitschrift für Geopolitik*, edited by the interwar geopolitician Kurt Vowinckel and the sociologist Karl Heinz Pfeffer, both of whom had a Nazi past. In the same year both Albrecht Haushofer's book *Political Geography and Geopolitics*, written during the war (Haushofer 1951), and a new edition of Heinrich Schmitthenner's *Lebensräume im Kampf der Kulturen* ('Living spaces in the clash of civilisations'), originally published in 1938, came out with Vowinckel.[4] Schmitthenner's work is particularly interesting, given his unwillingness to abandon the term *Lebensraum*. In the opening pages of his second edition, he explained that whilst the world had undoubtedly changed since his 1938 edition, the fundamental questions and coordinates of world politics had not. He went on to claim that

> National Socialism may have misused the word Lebensraum, but the term itself is not to blame. Goethe, after all, already used it. Biology and Geography then gave it its scientific content. And to describe the word struggle [*Kampf*] as militaristic is foolish. Struggle is a fact of all life and one that has no ethical content (Schmitthenner 1951: 6).

In his defence of the concept of *Lebensraum*, he argued that Europe, a continent in demise, would join the process of 'national dying' (*Völkersterben*) if it did not seek new living space in Africa (Schmitthenner 1951: 210).[5] Whilst Schmitthenner would remain one of the few West German commentators to continue using the term *Lebensraum*, other geopolitical concepts and ideas were starting to be articulated in a more forceful manner in the reborn *Zeitschrift für Geopolitik*.

As the editors of the first issue of *ZfG* explained, the journal was to study 'geographical, biological, psychological, economic, social and ideological forces with political impact' (Pfeffer & Vowinckel 1951: 80). In the spirit of Cold War era area studies, the journal sought to provide its readership with articles on the geopolitics of various world regions. At the cost of 2.30 Deutschmarks, *ZfG* offered an eclectic mixture of geopolitical writing, featuring many North American scholars, some female voices and even a few Marxists. Under the heading *Freie Aussprache* ('free debate'), *ZfG* included a platform for unorthodox opinions, free from censorship. Whilst there were some who advocated Konrad Adenauer's policy of *Westbinding*, others continued to hold anti-Atlantic positions. But even those who hoped for a more independent German geopolitics generally saw Germany as part of 'Western civilisation' rather than as a power that looked both to East and West, as one observer noted (Schnitzer 1955: 415). Indeed, one of the key discontinuities with the interwar period was the relative absence of advocacy for a German–Soviet axis, which had been so dear to Karl Haushofer. In 1955, Schnitzler estimated the journal's circulation at around 3,000 (Schnitzler 1955: 423), which would have only been slightly less than the 4,000 copies it sold in 1928 (Hepple 2009: 388).

Key to the way in which *ZfG* legitimated the return of geopolitics was the endorsement of scholars from the other side of the Atlantic. The North American professor Felix Wassermann, for instance, wrote a piece about Karl Haushofer's oeuvre in 1952 that promised that the latter's legacy would be kept up by a new generation of geopoliticians. 'The misuse of geopolitics through a totalitarian state, both of the National Socialist and Soviet kind', he felt, did 'not give free peoples the right to ignore its lessons' (Wassermann 1952: 726; see also Greenwood 1952). In 1951, the editors thanked Karl Haushofer for his dedication to the *ZfG* and emphasised that the question of Germany's geographical position would remain of central importance to the journal's remit. In 1954, the editors explained that the journal still stood in the tradition of Halford Mackinder, Alfred Mahan and Karl Haushofer and therefore interested in the continuing dynamic of land and sea power, the awakening of Asia and the struggle for *Lebensraum* as the defining motivation for historical change (Pfeffer & Vowinckel 1954: 193).[6]

And yet, only five years after *ZfG* was republished, the renaissance of geopolitical thought started to run into difficulties. Admittedly, *ZfG* had been greeted with mockery from parts of the West German press from the outset (Der Spiegel 1951) and the journal also faced competition from another periodical, *Außenpolitik* ('Foreign Policy'), which had been set up by German diplomats and which featured articles by German observers as well as translations of articles by the likes of Arnold Toynbee, George F. Kennan and Hans Morgenthau. Whilst *Außenpolitik* did publish on vaguely geopolitical themes – from overpopulation and the distribution of natural resources to regional knowledge – it was more oriented towards the theories and concepts of political science than those of the geographical discipline. The major blow to *ZfG* came in 1956 when it was absorbed by a politics and sociology journal because of financial difficulties that had plagued it since its relaunch (Schöller 1957: 4). Now published under the title *Zeitschrift für Geopolitik in Gemeinschaft und Politik* ('Journal for Geopolitics in Community and Politics'), the revamped journal was described by the new editorial team as offering an explicit 'change of course' away from traditional geopolitics and towards a more comprehensive understanding of the political (no author 1956).[7]

It is difficult to miss the fact that the re-emergence of *ZfG*, which had been the key medium of interwar geopolitics, was a failure. Peter Schöller, whose work played an important role in critiquing the renaissance of geopolitics in West Germany during the 1950s (Sandner 2000), would argue in 1957 that the return of geopolitics after the war was no more than a 'meaningless rebloom' that failed to achieve much resonance because of the 'dilettante' and 'scientifically dubious' nature of its articles (Schöller 1957: 4).[8] One might respond to this that a similar lack of scientificity had hardly limited the appeal of interwar geopolitics in the first place. Moreover, by focusing on geopolitical writing by academics *only* (for a similar approach, see Michel 2016), Schöller was missing the point that a new geopolitics was now emerging not in universities but in military circles.

For what was noticeable in the first volumes of *ZfG* was the presence amongst the authors of a number of former Wehrmacht generals. Some of these military commentators were unrepentant Nazis. Bernhardt Ramcke, for instance used *ZfG* to express his unhappiness with the way in which former Wehrmacht generals were treated by the Allied powers (Ramcke 1952).[9] But whilst Ramcke had only very little, if any, influence on postwar German politics, other contributors to *ZfG* were much more influential. Heinz Guderian, mentioned in Chapter 2, was to advise the budding German ministry of defence ('Amt Blank') in the early 1950s. In 1951, he used an article in *Zeitschrift für Geopolitik* to argue that 'the problem of time and space within strategy was timeless' (Guderian 1951b: 7). Speaking again of the eternal interplay of continental and sea powers, he warned fervently of the dangers of Western demilitarisation after 1945. 'Largely unnoticed by the West German public', Gross (2016: 260) concludes, 'former high-ranking German officers were as early as January 1946 addressing the thinking of the Wehrmacht in cooperation with the Operational History (German) Section of the U.S. Army's Historical Division'. Indeed, 'the Wehrmacht's tactical and operational experience in the East quickly became the focus of American interest at the start of the Cold War' (ibid.).

At the same time as former Wehrmacht generals were clandestinely integrated into the knowledge–power complex of the North Atlantic Treaty Organisation (founded in 1949), they were continuing to publish in *Zeitschrift für Geopolitik*. Friedrich Ruge, a vice admiral during World War II and soon to be the commander of the postwar German navy, used the journal in 1955 to publish an article on the politics of sea power. Under the title 'The forgotten sea', he urged his audience to recognise that the Federal Republic had now joined the sea powers in their struggle against the Soviet land bloc (Ruge 1955a). He explained that it was not just the ideological opposition to the Soviet Union that had motivated Bonn to join the West but a new and better understanding of sea power itself. He emphasised that it was not the first time in history that Germany had been in an alliance with sea powers and that the navy would play an important task in the next war, given its relative immunity to nuclear warfare. 'The most pressing task for German politics', he argued, was 'to keep the German nation [Volk] alive' (Ruge 1955a: 357).[10] People like Ruge and Guderian may have chosen *ZfG* to float their ideas, but they were not dependent on the journal. Soon they would be joined by others in articulating a new German geopolitics.

## The Contours of a New German Geopolitics

Geopolitical debates in the 1950s and 1960s were in many ways like those of the interwar period – diverse. To speak of one geopolitical narrative would thus be to overstate the commonalities amongst different geopolitical positions that were voiced in the Federal Republic. Indeed, some anti-Atlantic positions published in *Zeitschrift für Geopolitik* were in direct opposition to West Germany's official policy of *Westbindung*. And yet, it is possible to identify the contours of a new geopolitical discourse that emerged in the early 1950s and which can be found intact until the 1960s. This discourse both replicated and inverted the geopolitical debates of the interwar period in important ways. Firstly, by adopting an anti-Soviet position, this new German geopolitics now valorised sea over land power and rejected neutrality in favour of *Westbindung*. Secondly, although it abandoned the goal of territorial conquest, it continued to be concerned with Germany's unfavourable geographical position and territorial size. Some writers even retained the conceptualisation of the state as an organism that had to struggle for living space, even though the latter was now deemed to be sought underground.

Perhaps the crudest form of this new discourse came from the former general Heinz Guderian whose 1950 book *Kann Westeuropa verteidigt werden?* ('Can Western Europe be defended?') started out by suggesting that an answer to this question could only be reached by taking 'a globe or a world map into one's hands'. This, he explained, would reveal that Europe, 'our continent, which we were so proud of, [was] but a modest marginal strip at the western rim of that gigantic landmass that consist[ed] of the Soviet Union, China, India and a number of other states' (Guderian 1950: 11). Featuring discussion of the influence of climate on the national character of nations, Guderian placed the current struggle against the Soviet Union in the context of 1,500 years of European history, claiming that:

> the German nation's geographical position has remained the same for millennia. The Germans always inhabited Europe's centre. [...] All larger movements of peoples [*Völkerbewegungen*] that emerged from Asia just to be poured out over Europe first hit the Germans who had to fend them off (Guderian 1950: 12).

With the help of a number of maps, Guderian sought to reveal that the 'East' had encroached further and further upon a nation, which despite being 'Europe's centrepiece', had been 'torn apart' and deprived of its 'vitality' (*Lebenskraft*) (Guderian 1950: 12). Similarly, his colleague Günther Blumentritt wrote in 1952 that '[t]ime and time again in the course of history, storms of peoples' (Volksstürme) had emerged 'from this cold space into Western and Southern Eurasia.' Their aim, he explained, was 'the acquisition of "warm" land, of ice free oceans and harbours' (Blumentritt 1952: 13).

Guderian saw the Soviet Union as having achieved a 'state of near autarky' and invincibility from nuclear attack, not least by moving its most strategically important arms production underground (Guderian 1950: 37). Although he was in favour of the newly founded North Atlantic Treaty Organisation, he was sceptical as to whether the alliance would be able to stop the Soviet attack in its current form. Crucially, Guderian argued that the western Allies had made the mistake of attacking the territories held by Germany in 1944, rather than allowing it to continue fighting what he saw as the main enemy of Western civilisation, namely the Soviet Union, 'a dictatorship compared to which Hitler's dictatorship was but a weak reflection' (Guderian 1950: 28). Surprisingly, the book was endorsed by the American High Commissioner for Occupied Germany, John J. McCloy, despite Guderian's unwillingness to distance himself from National Socialism (Searle 2003: 85). Other proponents of the new German geopolitics were more nuanced and did not speak favourably about the Third Reich. Nevertheless, they would argue along similar geopolitical lines.

Firstly, and most obviously, the new geopolitics was of course anti-Soviet and pro-Atlantic. The aforementioned Friedrich Ruge would argue in 1963 that the Soviet Union's 'imperialist' politics had indeed led to the annexation of 500,000 square kilometres and 25 million inhabitants and had moreover brought seven states of almost 1 million square metres and 87 million inhabitants under its control (Ruge 1963: 113). As Blumentritt argued,

> If there really was a Bolshevik attack on the West, then we would see Germany and Austria being flooded by the Russian-Asian masses and the [Soviet] satellites, which would have been coerced into joining in. We have got to know the Russian [sic] in two World Wars and know that he would have no consideration for Germany. With brutal force, he would deport millions of people to the East and force those left into becoming soldiers and labourers. [...] Our personal freedoms would be no longer, the peasants would have nothing in their farms, citizens would be faced by empty shops and teachers would not be able to teach freely in their schools. [...] Whether neutral or not, nobody would be able to duck out of this war. This is not about dogmas or juridical doctrine, it is not about international law or other fancy terms, it is about our bare life! In such a struggle for survival, there cannot be any compromise! (Blumentritt 1952: 42).

Indeed, Günther Blumentritt was another former World War II general who had become particularly influential in advising the budding new German ministry of defence (Searle 2003: 111). It was the insistence of men such as Blumentritt that the Bonn Republic would have to learn to secure itself by entering into an alliance with the United States that chimed well with Adenauer's policy of *Westbindung*. Blumentritt would in 1960 write a book under the title *Strategie und Taktik* ('Strategy and tactics') which attempted to be a history of strategy from 500 AD to the nuclear age. Arguing against the idea that it was impossible to use nuclear weapons, he held that only deterrence and a balance of terror would prevent an outbreak of nuclear war. In order to ensure peace, he argued, one simply had to observe the globe and the territorial boundaries that were marked on it. 'New technological advances', he felt, 'had shrunk distances' to such a degree that 'intercontinental missiles could now reach almost all places on earth in the shortest of timespans' (Blumentritt 1960: 153). Neutral states, he argued, had simply lost their significance as these missiles would fly over them just the same. We should remember here that it was Chancellor Adenauer who had continually rejected neutrality by claiming that radioactive clouds would not stop at the territorial borders of neutral states (Adenauer 1957). Blumentritt was also convinced that 'every thinking person' could make their minds up about world politics 'without too much reading'. 'Sober thinking', he felt, was enough, if combined with a close study of the globe.[11]

Secondly, as we have already noted, the new German geopolitics articulated a reformed understanding of the land and sea power dynamic. In his 1955 book

*Seemacht und Sicherheit* ('Sea power and security'), Friedrich Ruge argued that the Federal Republic had now joined the sea powers, even though the German public had not fully understood or even noticed this. Calling for a 'new Wehrmacht', he argued that only a new German army could give expression to the German nation's 'will to live' and to reunite territorially (Ruge 1955b: 66). He held that in order to achieve these goals, Bonn had to not just support offensive forms of deterrence but to embrace the idea of sea power itself. Similarly, Erhard Jagemann had divided Europe according to a *maritime* and a *continental* circle, the former incorporating England, Scandinavia and northern Germany, the latter made up of Russia. The maritime circle, he argued, stood for European individualism, the continental circle for Asiatic collectivism (Jagemann 1955: 5–6).[12] In the eyes of these observers, the transatlantic West was thus a 'community of fate' (*Schicksalsgemeinschaft*) in 'the new struggle amongst civilisations and races' for space (Schmitthenner 1951: 226). Hitler, they argued had simply underestimated the importance of sea power (Ruge 1955b: 20).[13]

This new love of sea power was evident in a 1957 edited collection *Seemacht heute* ('Sea power today'), which was endorsed with a preface by Adolf Heusinger, then Inspector General of the Bundeswehr and later Chairman of NATO (Figures 3.1 and 3.2). Heusinger had been an advisor to Hitler and had served as the Chief of the General Staff of the Wehrmacht during the time of the attack on the Soviet Union (Bald 2005: 50). The book's jacket explained the

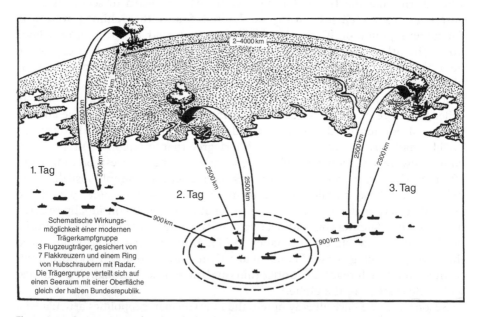

**Figure 3.1**   Representation of sea-launched nuclear war. Source: Reproduced from Ruge 1955b, p. 56.

**Figure 3.2**    Front cover of 1957 book that was endorsed by senior military figures. Source: Author.

need for such a text by stating that 'throughout history, continental thinking ha[d] often lead to doom'. In its pages, the former captain at sea Alfred Schulze-Hinrichs argued that both world wars had revealed 'that as a strong continental power, Germany had been defeated by the sea powers despite its sparkling victories at land' (Schulze-Hinrichs 1957: 33). He went on to conclude that sea power had itself proven a 'significant if not decisive moment in a great war' (ibid.). In the atomic age, sea power had the edge on land power due to the fact that targets at sea were much more difficult to take out than those on land (Blumentritt 1956: 7).

Thirdly, and much as during the interwar period, Germany was again understood as a fragile state whose geographical position exposed it to a whole range of military threats. 'After heavy losses in blood and property,' Ruge wrote, 'the German nation (*Volk*) lies weakened and divided in the earthquake zone between the two power blocs' (Ruge 1955b: 7). A few years later he added that 'Germany's position in the centre of the continent and its long boundaries, which are unprotected by nature, as well as its many neighbours, not all of which were always peaceful', demanded of the nation the acquisition of 'strong and responsive armed forces' (Ruge 1963: 37). Interestingly, this return of a focus on Germany's central position (*Mittellage*) was endorsed by strategists elsewhere, too. Ferdinand Otto Miksche, who held positions in different Allied armies during and after World War II, wrote that the 'well-known German conception that attack [was] the best defence [was] not, as some might think, born out of a typically German form of aggressiveness'. Rather, he claimed, it was 'conditioned through the country's central position' (Miksche 1955: 14). Erich Hampe, who had served in the Supreme High Command of the Wehrmacht and would become the Federal Agency for Civil Defence's first president (see also Stenck 2008: 78), asked in a similar vein:

> And today? Instead of a great empire (*Reich*) there is only a divided nation (*Volk*), one half of which is still under control of the Eastern victor. The Federal Republic has a long and bleeding border towards this victorious state. Its territory is open to its grasp. [...] These are hardly good aspects for the defence of the now liberated German soil [*Boden*] (Hampe 1956b: 25).

And yet, he continued, the situation was not as hopeless as it seemed, for this border was also the border of the 'free world against Communism'. The river Elbe, he explained, was where France, England and the United States were being defended, too.[14] Indeed, the 1966 civil defence survival guide *Die Kunst zu Überleben: Zivilverteidigung in der Bundesrepublik* ('The art of survival: Civil defence in the Federal Republic') explained that only Greece and Turkey were situated in a similarly exposed position by the iron curtain but warned that, unlike these states, the Federal Republic was much more densely populated (Kremer 1966: 11). The pamphlet's authors, endorsed by the Federal Agency for Civil Defence (Vulpius 1964: 17), worried that Germany's territorial compactness, 'which had always been apparent when observing a globe, would further shrink' (ibid.: 12).

Fourthly, it is possible to detect within the discourses of civil defence in particular a return of the term *Vernichtung* ('annihilation' or 'extermination'). As we have seen in Chapter 2, *Vernichtung* had already been a central concept for Ratzel and Haushofer, denoting the inherent nature of modern warfare itself. 'The objective of exterminating the enemy in the shortest time possible emerges directly for both East and West from the functioning of modern weapons systems', Kremer wrote (1966: 16). The Federal Republic had 'good chances' of partially surviving a war if it was progressively prepared for a limited nuclear war with tactical nuclear weapons and if the entire civilian population was brought in line with West Germany's civil defence policy (ibid.: 18).[15]

In true Ratzelian manner, civil defence planner Hampe spoke in his 1956 book *Im Spannungsfeld der Luftmächte* ('In the field of tension between the air powers') of a 'struggle to be or not to be' in which the Federal Republic's 'geographical location as a border people [*Grenzvolk*]' had placed it (Hampe 1956a: 10–12). Invoking the Schmittian language of sea, land and air power (Schmitt 1942[2001]), Hampe concluded that Europe had become no more than a larger air space in which the Federal Republic was particularly vulnerable. Hampe posed the question as to whether West Germany would remain a 'unitary and reactive organism' once the 'chaos of mass destruction' had been unleashed (Hampe 1956b: 14). As a 'nation [Volk] between the fronts', he felt, West Germany was threatened 'in its very existence' (Hampe 1956a: 41–2). This led him to conclude that it was imperative for Bonn to protect its population from the effects of tactical nuclear warfare. Any form of civil defence would have to 'serve the social, hygienic and humanitarian life of the population', he explained (ibid.: 72). In unmistakably biopolitical overtones, he held that every citizen would constitute itself as a 'living element of the state' who had to learn to protect itself (Hampe 1956a: 46). Indeed, in another publication he argued that

> Total attack is one side of the coin, the flip-side is total defence. Total defence comprises not just military but civil defence. The two belong together and form a larger whole. Thus contemporary war is Janus' faced in that one face looks outward towards the enemy, whilst the other looks inwards towards a nation's vital functions [*Lebensfunktionen*]. Only if the latter are sustained, will the fighting frontline possess resilience and military clout (Hampe 1956b: 9).

This chimed with Scharpff who reminded readers of *Zeitschrift für Geopolitik* in 1955 of Haushofer's way of conceptualising the state as an organism with administrative 'cells', which should be firmly embedded within the natural landscape. Similarly, Kremer wrote that the household is the 'lowest unit and cell of self-protection' (Kremer 1966: 46). It is interesting to note that the new German geopolitics would also make an appearance in the work of those who were neither associated with interwar geopolitics nor a part of the postwar military circles. The historian Ludwig Dehio, for instance, claimed in his 1955 book *Deutschland und die Weltpolitik im 20. Jahrhundert* ('Germany and world politics in the 20th century') that Prussia had managed to impose 'a new biological, spiritual and especially political vitality' upon the 'historically and culturally impoverished East' (Dehio 1955: 11). Posing the question as to whether in a dying system of states 'the geographical notion of Europe could now become an organism' (ibid.: 110), he answered that in order to emerge as such an 'organic entity', the otherwise 'dwarflike' (*verzwergt*) and 'nonviable' (*lebensunfähig*) states of Europe would have to merge in 'solidarity against Communism' (ibid.: 115). The chance for European unity arose not in the East, 'where Europe's essence [was] being crushed by Russia's brutal new ordering but in the West, where the Anglo-Saxons had been forced to reorganise what remain[ed] of Europe [*Resteuropa*] since the outbreak of the Cold War' (ibid.: 116).

Interestingly, *Zeitschrift für Geopolitik* only rarely touched upon questions of nuclear strategy in detail during its first five years. Elsewhere, however, proponents of the new German geopolitics were willing to absorb ideas from the emerging literature on nuclear strategy. Speaking at the annual meeting of the Association of the United States Army in 1960, Hans Speidel, a former Wehrmacht general and now the Supreme Commander of the NATO ground forces in Central Europe, argued for the crucial importance of West Germany's land forces in a prospective nuclear war with the Soviet Union. It was only through such forces that the 'aggressor could be defeated and exterminated and one's territory cleansed from the enemy', he argued (Speidel 1960: 168). He ended by stating in a biopolitical vein that the army remained 'the beating heart of the nation' (ibid.: 169) and that it was only within NATO that the German 'willingness to defend' could lead towards a 'better world'. Speidel would argue in 1969 that whilst massive retaliation meant national suicide for Germany, a selective use of nuclear weapons could ensure the nation's survival. Implicitly invoking Rudolf Kjellén's insistence that a state's territory was perhaps more important than its population, he argued that nuclear weapons should be used 'if significant loss of territory could not otherwise be prevented' (Speidel 1969: 67). It was the very same Speidel who allegedly participated in the preparation for the deportation of French Jews to Auschwitz in 1942 (Bald 2005: 50).

There had already been a sense amongst German generals in the early 1950s that the Cold War was but an extension of the Third Reich's military struggle against the Soviet Union. The ways in which they tied their geopolitical concepts to those of Anglo-American sea power gave the generals the possibility to contemplate and plan another war of annihilation without giving the impression that they were in any way still caught up in the intellectual and strategic universe of Nazi geopolitics. But what is also very noticeable is the way in which biopolitical ideas about population management as well as notions of the state as an organism managed to find their way back into the preparation for nuclear war.

## Return to the Soil

When the introduction of American nuclear weapons on West German territory first started prompting resistance in 1957, Chancellor Adenauer asked German scientists to speak out in favour of nuclear weapons (Cioc 1988: 80). In the same year, the nuclear physicist Pascual Jordan published his fantasy of an atomic Germany. He had been chosen by Adenauer as one of the few scientists willing to defend the government's stance. Whilst not explicitly written as a geopolitical or biopolitical pamphlet, traces of the German geopolitical tradition are very noticeable in its pages. Framing his arguments with the language of 'biologically necessary hereditary health [*Erbgesundheit*]', he argued that we should not assume that a nuclear catastrophe would necessarily also lead to 'extinction' (*Auslöschung*) (Jordan 1957: 176). There was simply no evidence, he claimed, that the effects of World War III would be graver than those of the previous world war. 'Even if a worse catastrophe were to reduce humanity to a mere fraction of its previous existence in the year 2050' (ibid.: 177), he held, 'then by 3050 the problem of overpopulation would have returned as the most burdensome of problems'. After all, the 'problem of overpopulation had not even been solved by two world wars' (ibid.: 177). In true Ratzelian manner, he concluded that human existence would remain endangered in the next millennium, just like that of the woolly mammoth or cave bear had been in the past. His solution was the development of 'underground cities', which were already made necessary because of the increase in traffic and the need for multistory roads, which could more efficiently be built in subterranea.

This idea of underground living space, it should be emphasised, was not intended by the prewar German geopoliticians at all. The young Ratzel had imagined a world of incredible wealth and fairytale creatures beneath the crust of the earth's surface (Ratzel 1905[1966]: 9) – but he died before the birth of the bunker. Haushofer did know about bunkers but was sceptical of them. He had been keen to stress the difference between a 'dead defence of space' (*toter Raumschutz*) and a preferred 'vital will to space' (*lebendiger Raumwille*), in which he opposed an inferior 'French' style of producing defensive architecture to a more active defence, with 'plane, tank and firearm' (Haushofer 1934b: 594). Expressing a critique of bunkered installations, he claimed that the answer to other nations' *lebendiger Raumwille* could not be made of 'steel and concrete', however useful those materials seemed (ibid.).

After 1945, there was a noticable concern with subterranean living space within the new German geopolitics. Ernst Samhaber, a journalist writing anxiously in *Zeitschrift für Geopolitik* about West Germany's vulnerable geopolitical position, recommended that suitable fortifications were built (Samhaber 1952: 654). Similarly, Friedrich Ruge argued that Germany's new armed forces had to be backed up with significant civil defence measures (Ruge 1963: 88). His colleague Leo Freiherr Geyr von Schweppenburg felt that West Germany should look to the East for inspiration in finding a way to go underground. A general during World War II and now an advisor to the US army, von Schweppenburg was concerned that Germany had become a 'buffer state between East and West and that such powerlessness had always invited invasion in the course of history' (von Schweppenburg 1952a: 81). In his 1952 book *The Great Question: Reflections on Soviet Power*, he argued that the historical transformations of the twentieth century had driven 'the Eurasian land- and population mass' towards becoming 'politically and economically superior' (von Schweppenburg 1952b: 21). Decorating his book with images of typical Russian racial types (*Volkstypen*), he wrote with great awe of the Soviet 'contempt for death' (*Todesverachtung*) and argued that a Russian soldier could endure much more than his Western counterpart:

> His hard composition and his affinity to nature make him the kind of combatant whose character emerges only fully when the nature of other and racially inferior [nations] drives them towards evasion or even failure. [...] The Russian is a kind of nocturnal animal, resourceful and without nerves. [...] He will disappear into the earth like a clever mole (ibid.).[16]

Von Schweppenburg was expressing a regard for the way in which the Soviet body itself was able to turn the earth into its servant by disappearing into it. It is thus perhaps no surprise that he spoke with great admiration of 'gigantic, secure – because subterranean' air bases that the Soviet Union had constructed in East Asia (von Schweppenburg 1952b: 34).

Erich Hampe, the Federal Agency for Civil Defence's first president, reminded his readers that it was in Hiroshima that 'subterranean shelter with considerable earthwork' had saved many lives (Hampe 1956a: 48). In his writing he included not only geopolitical maps that revealed what he deemed to be the fragility of West German territory in the upcoming war but also blueprints of civilian bunkers that would be able to ensure the survivability of the German nation in a nuclear war (see Figures 3.3 and 3.4). In Hampe's words, 'medieval cities sought to protect its citizens through the construction of high and insurmountable walls. Today, the protection of the population has shifted underground' (Hampe 1956a: 60).

We are reminded here again of Rudolf Kjellén's insistence that the nation-state should, 'if necessary', be able to survive 'behind closed doors' (Kjellén 1917: 162).

## Beyond the Taboo

Whilst the term *Geopolitik* was tabooed in mainstream politics and the subdiscipline of political geography was marginalised in German academia, the multiple discursive practices of geopolitics were not themselves discredited in the Bonn Republic. Unlike during the interwar period, however, the new German geopolitics could be found not so much in academic as in *military* circles. Amongst those who have touched upon the question of geopolitics after 1945, the role played by former Wehrmacht officers in giving birth to a new tradition of geopolitics has so far gone unnoticed. The new German geopolitics was not, as Kost (1988: 2) has suggested, an unmodified version of interwar geopolitics but one that absorbed new ideas from a variety of intellectual sources, even though the fundamental principles remained intact. Whilst geopolitics continued to be a niche discourse,

# Zeitberechnung für die Anflugzeit leichter Bomber und Fernwaffen

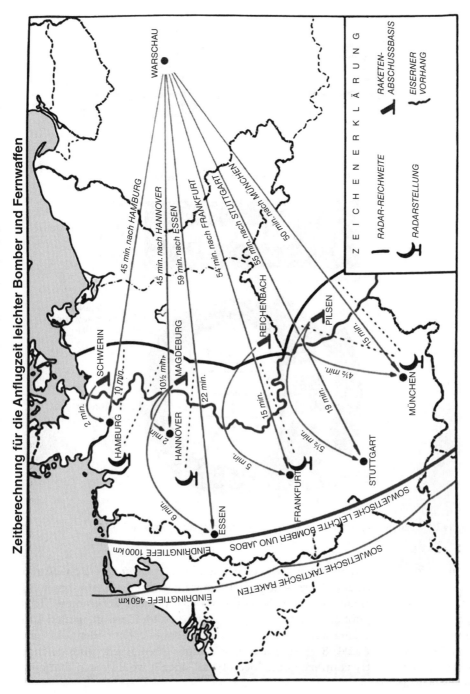

**Figure 3.3**    Map revealing West Germany's vulnerability to light bombers and long-range weapons. Source: Reproduced from Hampe (1956a: 57).

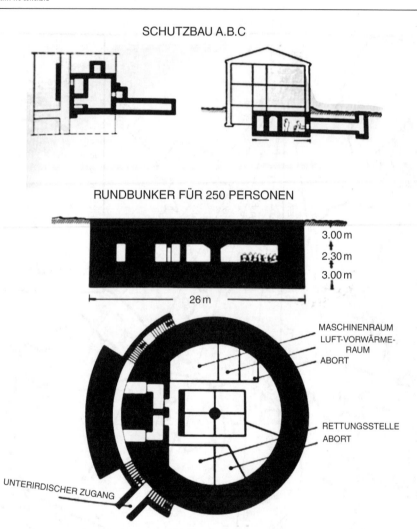

SCHUTZBAU A.B.C

RUNDBUNKER FÜR 250 PERSONEN

3.00 m
2.30 m
3.00 m

26 m

MASCHINENRAUM
LUFT-VORWÄRME-
RAUM
ABORT

RETTUNGSSTELLE
ABORT

UNTERIRDISCHER ZUGANG

**Figure 3.4**    Blueprint depicting nuclear bunker. Source: Reproduced from Hampe (1956a: 61).

some of its advocates, both those who contributed to *Zeitschrift für Geopolitik* and those who published elsewhere, were in fact senior figures in the establishment of the new West German army and the country's civil defence programme. This programme should not be thought of as purely *defensive*, for it was integrated into a more bellicose preparation for total war with a geopolitical opponent. As Carl von Clausewitz (1832[1984]: 357) had once famously remarked, 'pure defence would be completely contrary to the idea of war, since it would mean that only one side is waging it'.

The new geopolitics congealed around a number of familiar tropes, including the land and sea power dynamic and Germany's fragile centric position in Europe. As in the conceptual universe of prewar German geopolitics, political earth and political life were once again interwoven. In this new discourse, it was in particular the idea that the survival of the nation in a nuclear war might depend on building bunkers that distinguished it from its interwar predecessor. In this way, then, the bunker reconnected the biopolitical government of populations with the soil in which the nation was seen to be rooted. In many ways, the subterranean preparation for nuclear war seemed to materialise Ratzel's sense of 'an ever closer relationship between the nation as organism and the soil' (Ratzel 1941: 21).

It is important to understand that geopolitical thought was not so much promoted by surviving Weimar-era geographers but increasingly travelled across the Atlantic. Books, such as Miksche's 1955 *Atomwaffen und Streitkräfte* (original English title: 'Atomic Weapons and the Armed Forces'), later used in army academies for the training of West German officers, would, for instance, employ the term geopolitics quite self-evidently. Here the reader was told in a Ratzelian fashion that '[w]ars were always part of the lives of nations [*Völkerleben*], so that they seem to be somewhat of a biological necessity' (Miksche 1955: 192). In nuclear warfare, Miksche argued, there was no longer a significant distinction between frontline and hinterland. Under these circumstances, the whole nation had to be organised much like the military (ibid.: 178). Faced with the prospect of refugee flows out of the Central European battlefield, he explained, the state might have to resort to 'draconic measures' in preparing the combat area for battle (ibid.: 185).[17] It is perhaps no surprise that Robert McNamara would argue a few years later that a large-scale bunker construction programme was needed as a counterpart to the US ballistic missile programme (Monteyne 2011: 47, 75). Such hard-nosed endorsements from across the Atlantic set the scene for the new German geopolitics.

The former Wehrmacht generals who were so crucial in promoting the new German geopolitics may have been no more than a loose grouping in the mid-1950s, but one that had managed to find a common position, which centered on the assessment that Germany had left the rank of the great powers and that it had to realign with the West. The influence of the former generals on the emerging West German state was initially not direct but channelled through figures such as the British military writer Basil Liddell Hart. The latter had met many of the generals on a number of occasions in captivity and had turned into something of an advocate of them and their cause. In 1952, a year after the generals had published their book *Armee ohne Pathos* ('Army without Pathos'), which laid out the geopolitical foundations of a new German army (Weinstein 1951),[18] Liddell Hart had even met Adenauer to discuss the generals' concerns (Searle 1998: 346). German geopolitics had to be endorsed by the Anglo-American partners if it wanted to be heard. In the 1950s, the former generals provided unofficial advice, reports and also sat in on conferences (ibid.: 279). By the early 1960s, 12,360 former

Wehrmacht officers as well as 300 former members of the SS were in active service. By that time, West Germany had become a crucial member of NATO's conventional and even nuclear strategy (Bald 2005: 51).

In developing the argument that German geopolitics managed to survive its own funeral, we have so far avoided the question as to when and why geopolitics did eventually disappear as a more or less coherent set of propositions. Although this chapter has zoomed in on the early and mid-1950s, a period during which the new geopolitics crystallised in an unmistakable way, traces of geopolitics can be found until the late 1960s. It is in the 1960s, as Sprengel notes (1996: 36), that the term geopolitics 'trickles away', returning only to the fore in the 1980s with the *Historikerstreit* ('historians' quarrel'). The year 1968 marked not just the point at which *Zeitschrift für Geopolitik* would finally be discontinued, but it was also the year of the failed political revolution that would nevertheless come to profoundly reshape West German society and its relationship with its National Socialist past. A year later, a new Social Democrat-led government under Willi Brandt would abandon the crudest form of nuclear deterrence and reopen West German foreign policy to the East. Of course none of this meant that the Cold War and its logic of survival and extinction had itself disappeared from West Germany. On the contrary, the Federal Republic had become one of the most nuclearised territories on the planet. The new geopolitics certainly helped to legitimise the remilitarisation of a society that was otherwise wary of war. But most importantly, we can argue that geopolitics *could* disappear in the late 1960s because its aggressive architecture of bunkers and missile camps had now been established throughout the Federal Republic. The discursive superstructure of geopolitics was no longer vital in order to prop up the logic of survival and extinction that was materialised in the Cold War's military landscapes. It is to these landscapes, already prefigured in the revival of German geopolitics, that we will now turn.

## Endnotes

1   Ironically, by claiming that geography should be a resource of policy, Schmid seemed to valorise geopolitics in the very moment in which he rejected it as an overarching theory.
2   Farish (2004: 94) notes how Kennan was in the 1950s concerned not just with geopolitical space at a global scale but also at an urban one.
3   Morgenthau thought of geopolitics as a 'single factor' theory that elevated geography 'into an absolute that [was] supposed to determine the power, and hence the fate, of nations' (Morgenthau 1948[1993]: 174).
4   Carl Schmitt's *Nomos of the Earth* first appeared in 1950 (Schmitt 1950[1997]) and Otto Maull's *Politische Geographie*, originally published in 1925, was republished in 1956 in other publishing houses.
5   For a different attempt at salvaging the *Lebensraum* concept, see Maull (1956: 17).

6 Perhaps unsurprisingly, *Zeitschrift für Geopolitik* featured contributions from familiar faces. Carl Schmitt wrote an obituary (Schmitt 1952), Walter Christaller published a piece on subcontinents (1955) and Otto Maull wrote an article about the need to develop the European subcontinent into a political project (Maull 1951). It even featured a previously unpublished essay by Karl Haushofer on Indonesia and the South China Sea (Haushofer 1955).

7 As the journal started publishing on issues ranging from regional politics and airports to Mozart and Nietzsche, the remaining figures of interwar geopolitics abandoned the *ZfG*. It continued to be published until 1968, but geopolitics only played a marginal role in it and looked increasingly out of place next to the often politically rather esoteric work favoured by the editors.

8 Schöller (1957: 5) does go on to argue that geopolitics was 'not dead' in West Germany, but then limits his analysis to one book, Jagemann (1955).

9 Ramcke also took the opportunity to publish a very positive obituary for Marshall Petain in *ZfG* (Ramcke 1951).

10 *Zeitschrift für Geopolitik* debated at great length the emergence of the aircraft carrier (Pantenius 1952).

11 Blumentritt saw the globe as 'a speaking admonisher' that saved Germany from utopias. Many mistakes in Germany's past could have been avoided, he argued, 'if the responsible statesmen after Bismarck had stopped and looked at the globe every evening' (Blumentritt 1952: 12–13).

12 An important tension in his work and that of many of his contemporaries was that between a belief in the benefits of European integration and deterritorialising, on the one hand ('If the countries of Europe decided to tear down their borders, the continent could move towards a peaceful future'), and an insistence that the German territories that had been lost to Poland and the Soviet Union should be returned ('because German peasants had settled there hundreds of years ago'). Moreover, large parts of Europe were understood to be no more than 'the Occident's defence post against Asia' (Jagemann 1955: 75-77).

13 Similarly, Schmitthenner had argued a few years earlier that 'the earth ha[d] been carved into a continental communist and an oceanic non-communist' half. The Occident, Schmitthenner had felt, was doomed if it did not unite behind Western sea power to keep the aggressive East in check (Schmitthenner 1951: 210–11).

14 More than a decade later, the question of the geographical centre was still at the heart of debates on German foreign and security policy. In 1969, the *Deutsche Gesellschaft für Auswärtige Politik* (German Council on Foreign Relations), which remains until this day one of the most important think-tanks in Germany, published a study on 'middle powers' in world politics, which it defined as states that function at a meso level in the world system, in between the great powers on the one hand, and more local actors, on the other. Concluding that the Federal Republic had since 1945 become just such a middle power that was now able to shape politics at a regional level, the study remained preoccupied with the markers of geopolitical power: territory, security and population. Speculating about the role that territorial space continued to play in the nuclear age, it argued in a familiar geopolitical manner that '[v]ast space brings with it considerable advantages [...], it reduces the vulnerability to nuclear attack, it allows a state's economy and population to expand faster [...] and it communicates a fundamental sense of national greatness and significance that serves as the indispensable

psychological basis for world power [*Weltmacht*]' (Deutsche Gesellschaft für Auswärtige Politik 1969: 147).

15 Only in 1954 did nuclear weapons start to be considered tactical rather than strategic in US military thinking. It was a year later that the Federal Republic's military and political elites 'fully realised the extent to which the alliance planned to rely on nuclear weapons in the future' (Gross 2016: 277).

16 Von Schweppenburg concluded that it was important to recognise that Russians were naturally prone towards the idea of authority. Much like George F. Kennan, he argued that the 'indolently passive nature' of the Greater Russians could only be held together with 'brutal methods' (von Schweppenburg 1952b: 44).

17 Miksche also argued that 'under the pressure of a tremendous "pan-Slavic empire", reaching from the Elbe and the Danube all the way to the Pacific Ocean, what remains of Europe [*Resteuropa*] ha[d] only very modest chances of conserving its independence' (Miksche 1955: 190).

18 The book synthesised many of the ideas discussed in the new German geopolitics by bringing together 16 former Wehrmacht officers to reflect on West Germany's strategic position. It concluded that the postwar epoch was one in which Germany had to abandon the fantasy of ruling over Central Europe for the idea of being an ordering power [*Ordnungsmacht*] (Weinstein 1951: 157).

# Chapter Four
# Nuclear Living Space

*In the history of civilisation, we can observe an ever closer relationship between the nation as organism and the soil*

*Friedrich Ratzel 1882[1909]: 126*

## Überlebensraum

Roughly 16 miles south of the former West German capital of Bonn lies a remarkable relic of the Cold War. Surrounded by vineyards, forests and small half-timbered houses are the remains of the West German government's now abandoned nuclear retreat. Completed in 1972 and decommissioned in 1997, the bunker remains to this day the costliest architectural project that the Federal Republic has undertaken. Secured behind tons of reinforced concrete, heavy blast doors and a sophisticated filter system, 3,000 politicians, bureaucrats and military staff could have survived for up to a month. Like a hibernating organism, or so it was hoped, the state would have been able to endure the nation's darkest hour autarkically – buried in the very German soil about which Ratzel and Haushofer had obsessed in their writings.

Like the new capital city, the West German government's bunker in Marienthal is located in the Rhineland region – the same region that Halford Mackinder had once described as 'the most significant division in Europe' (Mackinder 1908: 2–3).

*Cryptic Concrete: A Subterranean Journey Into Cold War Germany*, First Edition. Ian Klinke.
© 2018 John Wiley & Sons Ltd. Published 2018 by John Wiley & Sons Ltd.

Comparing the Rhine in its geopolitical significance to the Indus and the Yellow River, Karl Haushofer had once warned that the overpopulated Rhenish living space would invite foreign domination if Germany did not find its own assertive geopolitics (Haushofer 1928: 8). Geologically and geopolitically, he claimed, the Rhineland's history had always been one of 'combat and struggle'. 'Nature' simply wanted 'no pacifism here' (ibid.: 6). Of course, Haushofer was ultimately proven wrong for in the 1950s and 1960s, the Rhineland would progressively lose its geopolitical relevance (Klinke & Perombelon 2015). And yet, a concern with geopolitical space lived on in the region. For in the thermonuclear age, it was no longer primarily Friedrich Ratzel's open spaces that were life sustaining (Ratzel 1901) but the confined living space of the nuclear bunker.

For the German geopoliticians, *Raum* had always been a 'supernatural force', which had 'framed the state, []composed the state, in the final analysis actually created the state' (Murphy 1997: 26). If we translate *Raum* into the English as 'space' or 'area', we may easily lose sight of the thick layers that the term has in the German language and the mystical qualities it has within the German geopolitical tradition. But *Raum* does not just denote the abstract space of a geopolitical map – the violent cartographic fantasy of *Lebensraum* – but also captures the English meaning of 'room'. Indeed, *Schutzraum* (literally 'space of protection') was the West German civil defence planners' preferred term for air-raid shelter during the Cold War. As a biopolitical space, the bunker seemed to give material expression both to Rudolf Kjellén's insistence in *Der Staat als Lebensform* that a nation should 'if necessary' be able to survive autonomously – 'behind closed doors' (Kjellén 1917: 162) – and Friedrich Ratzel's idea of 'an ever closer relationship between the nation as organism and the soil' (Ratzel 1882[1909]: 126). As we will see below, it is in its subterranean survival pod that the Cold War state managed to achieve that close relationship with *terra*.

By the mid-1950s, a heated debate had emerged amongst West German civil defence planners as to the way forward for the young republic, threatened as it lay at the heart of the European continent. One observer, Alexander Löfken, who would later become the head of the Federal Agency for Technical Relief (*Technisches Hilfswerk*), called for a new era of spatial planning to respond to West Germany's new political situation. With significant geopolitical undertones, he noted that '[a]s far as we can look back in history, we can always observe the eternal law of self-preservation' (Löfken 1954: 33). History, Löfken argued, was 'but an endless chain of struggles, of attack and defence' (ibid.). From the Roman limes and China's Great Wall to the Maginot line and the German Westwall (Siegfried line), construction had always been a part of defence measures, the civil defence planner explained. West Germany was in a particularly precarious position and had to re-evaluate its 'space'. This was because Germany's shrunken territory and its frontline position meant that the nation's 'remaining living space' would necessarily be turned into an area of operations in the coming war. And yet, rather than arguing for territorial revisionism, Löfken called for a new era of greater spatial planning (*Großraumplanung*) that targeted not the conquest and control of territory but rather the built environment. Germany would only manage to secure its *Lebensraum*, Löfken held, if it prepared its built environment for total war. No single building, he argued, should be built without air-raid protection. Only if civil defence was subordinated to military strategy, could Germany's 'living space of the future' be secured (Löfken 1954 34; see also Löfken 1960).

Whilst Löfken was unusual in continuing to use the term *Lebensraum* in postwar German civil defence planning until the 1960s, he was not alone in rethinking civil defence along geo- and biopolitical lines. We have already seen in Chapter 3 how Erich Hampe, the Federal Agency for Civil Defence's first president, kept on insisting that the state was an organism that struggled for survival in a competitive environment. But rather than thinking of *Lebensraum* in a strict Haushoferian sense as two-dimensional territory, civil defence planners like Löfken and Hampe became more and more concerned with the vertical dimension of living space. For in civil defence and elsewhere, subterranea was increasingly seen as the solution to the Federal Republic's geopolitical situation. Like the National Socialists before them, they would become obsessed with spaces of survival – quite literally *Überlebensraum*.

This chapter takes a closer look at the bunker as an architectural solution to West Germany's Cold War. It does so in two ways. Firstly, it seeks to reveal how the ideas popular amongst former World War II generals, outlined in Chapter 3, were from the 1950s onwards finding their way into the realm of civil defence planning. Of course, civil defence meant the preparation of homes and individuals for nuclear war, but it also meant that *Lebensraum* was increasingly sought in the vertical dimension. It could be secured, or so the Bonn Republic's civil defence planners hoped, through a large-scale public bunker construction programme. Secondly, the chapter will explore in more detail one particular subterranean structure, a bunker built not so much for the protection of the population than for the survival of the state's organs in total war – the West German government's retreat near Bonn. This structure had in fact been built *into* a former World War II underground slave labour camp, a subcamp of the infamous Buchenwald. Through a discussion of this bunker's technical and security features, we will examine the ways in which it inverted its predecessor spatially.

## Civil Defence and the Return of the Bunker

Cold War Civil defence programmes were of course not limited to West Germany but emerged in a large number of states during the 1950s and 1960s, both east and west of the iron curtain (Davis 2007; Berger Ziauddin 2016; Geist 2012; Monteyne 2011). As elsewhere, the key aims of West German civil defence were the maintenance of the state's ability to act in war, the protection of civilian life, the defence of vital resources and supplies as well as the support of military operations. Whilst civil defence would increasingly become concerned with threats that arose in times of peace, including accidents in nuclear reactors and natural disasters, it was originally concerned with war. Growing directly out of the Third Reich's air-raid protection schemes, West German civil defence was in the early postwar years preoccupied with conventional aerial warfare, but would soon become targeted at the effects of nuclear war.

Much as in other states, the Federal Republic's civil defence programmes incorporated both centralised measures taken by state institutions to prepare society for war, activities by a number of non-state organisations (such as the Red Cross and the Samaritans), and more decentralised forms of civic participation. The key institution for civil defence in West Germany was the Federal Agency for Civil Defence (*Bundesamt für zivilen Bevölkerungsschutz*)[1] which coordinated the interface between state and society in the preparation for nuclear war. The office had a wide-ranging set of competencies, including the construction and maintenance of sirens and other critical infrastructure as well as the production of guidelines for individual self-protection and the preservation of cultural objects. Most importantly, however, it was an advocate and coordinator of the nation's attempts to move underground.

Whilst the Cold War is often seen as a period of secrecy, civil defence had in fact to be public and inclusive. By the late 1950s the Federal Republic's civil defence programme was aided by around 80,000 volunteers (Lennartz 1958). Moreover, civil defence planners organised the entire population along a number of spatial scales, from the family to communities (125 inhabitants), blocks (500–1,000 inhabitants), districts (5,000 inhabitants), segments (20,000 inhabitants), sections (100,000 inhabitants) and sectors (500,000 inhabitants). At the largest scale, the Federal Republic was divided into ten warning areas, which loosely corresponded to the boundaries of the federal states. Whilst civil defence at the higher scales was organised by the state directly, inhabitants themselves had to coordinate civil defence measures at a micro level. At the smallest scale of the civil defence apparatus stood the individual home, which was based on the heteronormative assumption of the family unit. Echoing the US government's 'stay at home policy', the Federal Ministry of the Interior explained that the individual's security was 'better guaranteed in a place that is familiar to you' (Bundesminister des Innern 1964: 7).

By the mid-1950s, urban planners and engineers were starting to develop an interest in subterranea in a number of registers. Firstly, inner cities were commonly seen to have become architecturally saturated spaces, reeling under the growing traffic volume of the postwar era. Pointing to advances in the East bloc, particularly in Moscow and Budapest, one expert argued that public transport would have to completely 'disappear into the earth'. 'The third dimension is the only way left', and it was precisely this third dimension that now 'had to be conquered' (Klingmüller 1955: 199–200). Secondly, high-rise bunkers (*Hochbunker*), which had been popular in the Third Reich, were criticised as being too vulnerable in the nuclear age (Soxhlet 1953). Instead, 'natural tunnels and subterranean caves' should be converted into bunkers in times of peace so they could house industrial production, natural resources and people in wartime (Paetsch 1952: 10). Unlike the Atlantic Wall with its iconic above-earth structures, the bunker of the future would have to be invisible from plain view, hidden in the earth (Vanderbilt 2002). The state had thus to establish that ever closer relationship between the nation as organism on the one hand, and the soil on the other, the very link that had once preoccupied Friedrich Ratzel (Figure 4.1).

In principle, nuclear bunkers were not all that different from World War II air-raid shelters. They featured concrete casings that provided (some) protection from blast waves and radioactivity. Like the air-raid shelters, it was crucial for these nuclear bunkers to have a number of emergency exits in case the main entrance was destroyed or covered with rubble (Leutz 1956). On the inside, public nuclear bunkers were Spartan places, which featured no more than the bare minimum for personal comfort and survival: water and food supplies, seating and bunks, overhead luggage racks, air-raid suitcases (for clothes, cutlery, toiletries and some personal belongings), medical supplies, portable emergency toilets, and basic fire protection technology (Zimmermann 1958: 138). Emergency lighting and luminescent paint would guide the occupants around the bunker in the case of a power cut.

Whilst the Federal Republic's civil defence planners had initially tried to reuse and modernise old World War II bunkers, from 1961 onwards they increasingly turned towards new multipurpose facilities (*Mehrzweckanlagen*), which blended the bunker into the urban fabric (Bundesinnenministerium 1963: 5).[2] Rather than building structures that had as their sole purpose the protection of the population, they would look to construct bunkers within sites that were primarily built for other purposes, such as underground stations (see Figure 4.2), hotels and carparks. By 1972, 23 such multipurpose facilities had been built, nine were under construction and a further ten were planned (Bundesinnenministerium 1972: 78). Interestingly, such sites were increasingly constructed with the ruined building in mind (Bundesschatzminister 1968: 2), a practice amongst engineers and planners that became ever more popular during the 1960s (see Figure 4.3). For bunkers to be effective they had to have emergency exits that came out beyond the house's rubble. Much as with Albert Speer's theory of ruin value, the individual building was thus seen from a position in the future in which the nation had been wrestled down by the forces of history.

Much of the new concern with bunkers was framed through very familiar geopolitical discourses. Erich Hampe's Federal Agency for Civil Defence would frequently explain the need for civil defence through reference to West Germany's 'geographical position' (Bundesamt für zivilen Bevölkerungsschutz 1965: 4; 1970: 21). Of particular concern was the state's 'narrow territory' and its 'population density' (Bundesinnenministerium 1972: 15). Until the 1960s, the term 'defence

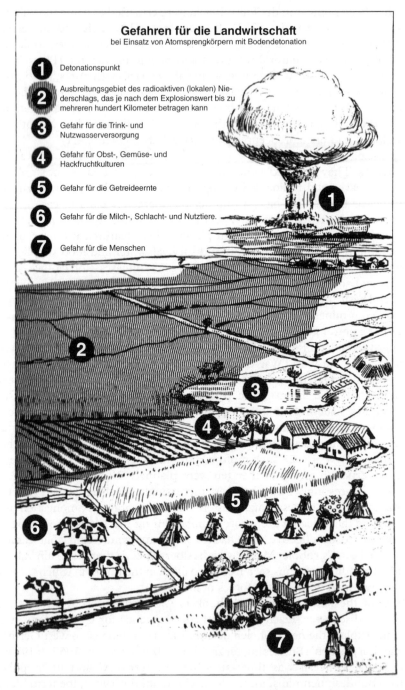

**Figure 4.1**    Effects of nuclear war on German agriculture. Source: Reproduced from Kremer (1963).

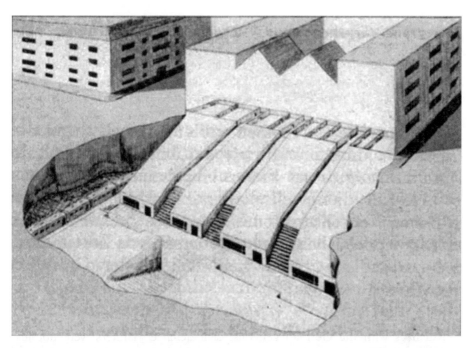

**Figure 4.2** Underground station functioning as a bunker. Source: Reproduced from *Ziviler Luftschutz* 6/1960.

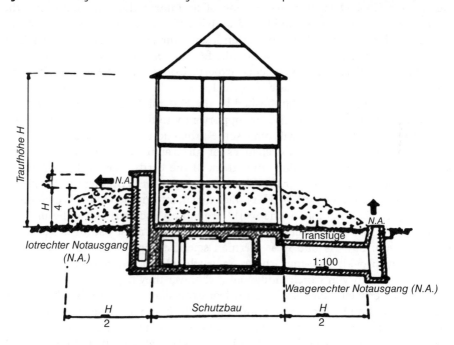

*Notausstiege und Trümmerbereich eines Mietshauses*

**Figure 4.3** Plan of a bunkered house in two states: before and after the blast wave. Source: Reproduced from *Atomgefahren: Was stimmt? Was kommt? Was tun?* (1962, p. 101).

geography' (*Wehrgeographie*), which had been employed by Haushofer during the interwar years, was used in the pages of *Ziviler Luftschutz*, a key forum for debates on civil defence.[3] Published by the Federal Agency for Civil Defence, this monthly magazine gave a home to insights from engineers, bureaucrats and scientists as well as printing reports on the state of civil defence in other countries. Alongside analysis of technological advances in the field of weapons technology and bunker construction, it also featured basic analyses of Cold War geopolitics. A 1960 article by Eduard Beyer, for instance, reminded the civil defence community that West Germany's geographical position remained unchanged and that neutrality was pointless given how the 'free world' continued to face an 'imperial power block' in the East (Beyer 1960: 141). Contrary to the prewar German geopolitical tradition, Beyer argued that West German defence policy would have to abandon the fantasy of autarky in the fields of the economy and security and concentrate instead only on naked survival, which he defined as 'the resistance of the whole nation and in its socio-political order against the threat to its life' (ibid.: 146). Proposing to organise the state according to a 'cell structure', he argued that civil defence would always have to be subordinate to military operations.[4]

Much as with the German generals, civil defence planners were not just looking to the West for inspiration. It was with great admiration that the Federal Agency for Civil Defence noted in one of its information brochures that the Soviet Union had already established a civil defence service that incorporated over 20 million of its inhabitants (Bundesamt für zivilen Bevölkerungsschutz 1961b: 6). Similar to the way in which the former generals had framed the earth as a place of salvation, civil defence experts would read the soil as the only resource for survival in the case of a nuclear attack. 'If no air raid shelters are available', a training manual published by the Federal Agency for Civil Defence advised, it was 'important to use any opportunity to find cover'. Even 'the naturally grown soil, the earth, will offer some protection'. The manual went on to explain that in order to increase chances for survival, the body had to be brought into a foetal position to endure the attack in the earth – crossed arms, hands under the armpits, face under the elbow and closed eyes (Bundesamt für zivilen Bevölkerungsschutz 1965: 12). 'Here, face down, internalised in one's own mind, and completely vulnerable to the world around', Masco writes, 'is the ultimate Cold war posture – a sightless, private bunker of the most pathos-driven kind' (Masco 2009: 25). As one West German scientist explained, the state should increasingly provide one-man 'holes in earth' (*Erdlöcher*) across the Federal Republic's territory in which individuals, wherever they happened to find themselves in the crucial seconds after a nuclear explosion, could survive the nuclear blast wave by simply disappearing into the earth. If an individual was more than 1 km away from the epicentre of a nuclear explosion, he argued, such holes were the best means of survival (Dählmann 1953: 169; see also Kremer 1963). Indeed, the earth was advertised by the Federal Agency for Civil Defence as a much safer material against radiation than poured concrete and brick (Figure 4.4).

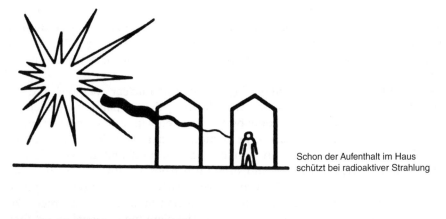

Schon der Aufenthalt im Haus
schützt bei radioaktiver Strahlung

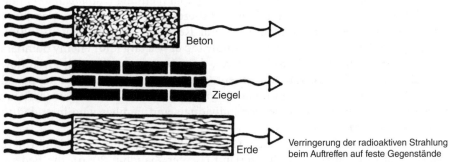

Beton

Ziegel

Erde    Verringerung der radioaktiven Strahlung
beim Auftreffen auf feste Gegenstände

**Figure 4.4**    The effects of radiation on concrete, brick and earth. Reproduced from Bundesamt für zivilen Bevölkerungsschutz (1961a, p. 7).

Clearly, the earth played a crucial role within the survivalist mythology of the West German Cold War. The subterranean bunker seemed inevitable in an age in which '[o]nly autarchic powers would be able to fight a long war' (Guderian 1951a: 26). In an era in which Ratzel's fantasy of controlling large open spaces seemed unattainable, the bunker would serve as the smallest possible contraction of a state, the smallest possible *Lebensraum*. Although the West Germans were obsessed with the possibility of subterranean survival, funds for civil defence were in fact limited in a society that was still recovering from the last war. At the same time, motivation to invest in civil defence measures clashed with a more general amnesia, a society-wide attempt to try and forget the experience of aerial bombardment. As one civil defence expert noted, 'the German population has lived through so much during the aerial bombardments of World War II that it does not want to be reminded of this five-year period of suffering' (Schmidle 1959: 254). We will return to this collective amnesia in Chapter 6. Thus, the efforts to push the Federal Republic into a bunkered existence had always to remain a fantasy. No more than a fraction of West Germany's population would have found shelter in public bunkers. Crucially, there had always been doubts as to the degree to which such bunkers in urban areas were actually able to offer protection, given how cities were themselves likely primary targets in the coming war.

Faced with both the ever increasing potency of nuclear weapons and an increasing fatigue around issues of civil defence, the parliament decided in 1965 that public shelters would only be constructed with a basic protection in mind rather than a more expensive protection for near hits. Indeed, by 1970, West Germany was spending on bunkers only around half of what it was spending on its warning systems and siren infrastructure (Bundesamt für zivilen Bevölkerungsschutz 1970: 7). From a very early stage, civil defence planners had promoted two alternatives to large civilian bunkers. Firstly, men like Erich Hampe had called for a new form of spatial planning (*Raumplanung*) that would disperse urban agglomerations and thereby spread the population more equally across the territory of the Federal Republic (see also Farish 2004). Densely populated urban areas were simply too easy a target for enemy bombers and missiles (Hampe 1952: 6).[5] And yet, this idea never appealed, given how difficult it would have been to implement. Another solution, much as happened in the United States, was to stimulate the private bunker construction economy by providing financial incentives for individuals and families who intended to build their own fallout shelters.

In the absence of a large-scale civilian bunker programme that could have rivalled that of countries like Switzerland (Berger Ziauddin 2016), the Federal Agency for Civil Defence thus turned in the 1960s to a strategy of urging the West German population to build shelters on a voluntary basis (Bundesamt für zivilen Bevölkerungsschutz 1961a: 10). Erich Hampe had already advocated the construction of such home bunkers in 1952, even though, at the time, he admitted that the funds for such an undertaking were not yet available (Hampe 1952: 5). In this vein, civilians were told to think 'geographically' about their surroundings. 'If your apartment lies in between hills or in a deeper valley, then you already possess a "natural air-raid shelter [Schutzraum]"' (no author 1962: 93). By the early 1970s, the Federal Republic was granting subsidies for private bunker constructions as long as they abided by certain guidelines (Bundesinnenministerium 1972: 77).

Interestingly, at this smallest of scales, civil defence was to be held together not so much by the patriarch of the family than by women (see also Masco 2008: 374). Urging Germans to emulate the gender divisions cultivated in American civil defence, German civil defence experts explained that 'the woman's civil defence begins at home' and that women thus had to be made familiar with the 'basic principles of civil defence' (Schützsack 1955: 121). It was a woman's responsibility to keep the home's fallout shelter, if it had one, functional and free from fire hazards. This was done by tidying and cleaning but also by making sure that the electrical equipment was up to date and in good working order. She should, moreover, be trained in first aid. Before securing the family's material belongings, she had to ensure the family's physical security. Being part of the 'nuclear family' (Farish 2004: 95) came with responsibilities. The population was urged to acquire knowledge of the surrounding landscape, including important public buildings, road and resource infrastructure, hospitals and chemists as well as the location of public bunkers themselves. Moreover, everyone had to be aware of the basic sirens for different types of attack, conventional and ABC (atomic, biological and chemical) as well as the 'all clear' signal after which it was safe for the nuclear family to crawl out of its bunker.

The attempt to control social relations by targeting the home also bled into more general attempts to regulate social behaviour in public bunkers. Much of this happened in a biopolitical register. Training manuals for civil defence personnel would routinely emphasise the need for hygiene and good manners in the bunker. 'Comradery, strict self-discipline, helpfulness, tidiness, cleanliness and punctuality', they explained, 'made bunker life easier' (Bundesamt für zivilen Bevölkerungsschutz 1965: 15).

> Personal hygiene is particularly important. Cold water should be used, as it hardens the body. Uncleanliness is repulsive and decreases the resilience against pathogens. Bodily excretions produce a vile body odour. Thus it is imperative to wash one's body thoroughly (including the feet) in the mornings and evenings and clean one's teeth. Finger- and toenails are to be kept short and are to be treated with a nail cleaner (not with a knife). Hair is to be groomed and beards should be shaved in the mornings (ibid.)

Reinforcing Foucault's insistence that biopolitics operates both at the level of the body and the population, guidelines for life underground echoed some of the hygienic regulations for soldiers in the battlefield. The latter, however, had also to abide by certain sexual standards, such as the imperative to keep away from 'women whose life choices point towards the possibility of sexually transmitted diseases' (ZDv 49/20 1961: 190). We are reminded here of course of Foucault's discussion of biopolitics as the regulation of, amongst other realms, sexuality (Foucault 1978: 145). Guides for Bundeswehr soldiers in nuclear war would go to great length to describe the bodily effects of nuclear war, from blinding to burns, and the effects of radiation sickness, both short and long term. Yet, the imperative was clear. 'If, after an atomic detonation, you experience symptoms of radiation sickness, then overcome your weakness, continue your service and help your comrades' (ZDv 49/20 1961: 149). Again, the earth was crucial as a space for protection and perhaps for dying. For the soldiers were both advised to seek protection from a nuclear blast wave in earth holes (Figure 4.5) and drag wounded comrades into such holes.

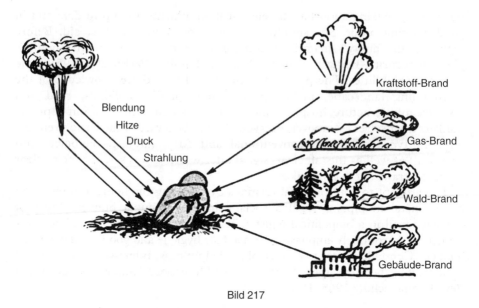

Blendung
Hitze
Druck
Strahlung

Kraftstoff-Brand

Gas-Brand

Wald-Brand

Gebäude-Brand

Bild 217

**Figure 4.5**    Guidelines for soldiers in nuclear war. Reproduced from ZDv 49/20 (1961).

But death was not just rationalised on the battlefront. There was a real danger that civilian bunkers could themselves necessitate a biopolitical decision. As the places in civilian bunkers were severely limited, it would have been necessary for an individual bunker commander to decide when the bunker was full and the doors had to be closed. Not only might children be separated from their parents, but some people might be squashed to death when the bunker's remote-controlled doors were closed (Westfälische Rundschau 1961). Often, this moral problem was solved architecturally by having the door operated by someone who could not see what was going on at the blast door. The door of the bunker thus functioned as an Agambenian threshold at which some were brought into the protective space of the bunker whilst others were abandoned to the nuclear holocaust outside.

Whilst the West German state was clearly preoccupied with the resilience of its soldiers and civilian population in the case of a nuclear war, it was also – perhaps even more – concerned with its own survival. From a very early stage, internal documents from the Ministry of the Interior show that there were concerns that a single nuclear weapon dropped on the new capital Bonn 'would be likely to annihilate the vast majority of ministries' in an instance (Bundesinnenministerium 1950). Soon, the ministry would be actively searching for its own subterranean retreat.

## From Camp to Bunker

In 1955, the Ministry of the Interior assumed the following scenario:

> War will begin with a period of the hardest nuclear attack. This period will last for around 30 days. The ability to defeat the enemy will depend on the capacity to survive these first 30 days of the war and to achieve military superiority. [...] The targets of attacks with thermonuclear weapons will primarily be Allied production facilities, nuclear weapon sites, Allied government centres, industrial and transport facilities that serve defence purposes, bigger ports and densely populated areas, the destruction of which will entail heavy losses for the conduct of war, particularly by undermining the population's morale (Bundesinnenminsterium 1955).

In line with NATO policy, the government's main aims during a nuclear war were the maintenance of (in the following order): 1) government power; 2) the operability of the armed forces; 3) the social order; 4) civil contingency planning; and 5) foreign policy (Bundesinnenministerium 1960b). In order to ensure the top priority, it was necessary to build a nuclear bunker that could ensure the survival of all vital ministries. After some discussion, a decision was taken to search for a site at some distance of the capital and to construct this shelter underground rather than above ground. Erich Hampe, the geopolitician and civil defence chief, put his signature on the order to find a site for the government's nuclear retreat. In the end, a decision was reached to reuse a site that had already been used during World War II. In fact, the story of this bunker had begun in 1913.

On the eve of World War I, the last German Kaiser gave the order to build a new railway line along the river Ahr in preparation for a surprise attack on France. As the railway was not finalised in time for the war, the infamous Schlieffen plan had to be executed without one of its supply lines. After the war, the site lay abandoned and slowly lost its strategic importance due to technological advances in the field of mechanised warfare and aviation – but the many tunnels, some of which were carved deep into the hills, remained. In 1936, after the National-Socialists had taken power in Germany, these tunnels were briefly used to grow mushrooms, which like the wine that ripened on the valley's slopes, was meant to rival superior French production (Janta, Rieck & Riemenschneider 1989: 140). When during World War II the German industrial centres became the target of heavy aerial bombing, production was often moved underground. In 1943, the railway tunnels on the river Ahr were turned into a small subterranean slave labour camp of around 200 inmates.

Poetically named Rebstock (grapevine), this subcamp of the notorious Buchenwald concentration camp was a production site for the Third Reich's cruise missile (V1) and the ground vehicles for the world's first ballistic missile (the V2). The camp reflected in function and design the much larger Mittelbau-Dora in Thuringia. Camp Rebstock was subject to a continuous influx and outflow of inmates as those deemed unfit to work were sent back to Buchenwald (Jungbluth 2000: 61). Towards the end of the war, Rebstock had become the target of British bombing campaigns. During the winter of 1944–45 in particular, the local population sought refuge in the tunnel and production was brought to a halt. It was in those days that the civilian population came into close contact with the camp inmates, as local residents have since testified (Gückelhorn 2002: 65), thus blurring the boundary between the camp's inside and outside. Whilst photographs of the camp show the archetypal barracks located in front of the reinforced railway tunnels, the camp's remains were unmarked after 1945, known only to the survivors and the local population. Rebstock first came into the public consciousness in the mid-1980s when an investigative journalist started exploring what he believed to be the West German government's wartime retreat and stumbled on stories of a World War II concentration camp (Preute 1984). In 1945, the Allies had destroyed the reinforced entrances with targeted explosions, as if to disallow geo- and biopolitics to return to this particular space (Bundesinnenminsterium 1959).

And yet, in the 1950s the site became the target of the government's preparations for the next war, a nuclear war that threatened to be fought on German soil. It was in this context that sovereign power returned to a valley whose inhabitants were already schooled in keeping subterranean secrets, as the West German state began turning the reinforced tunnels into a small underground city. Whilst the budget for the bunker was initially set at 43 million Deutschmarks, the costs had soon more than tripled. Even the government acknowledged in internal documents that the costs were spiralling out of control, particularly given that civilian bunkers at the time only had room for a fraction of the West German population (Bundesinnenministerium 1960a). And yet, this did not stop the government pushing on with the construction of its autarkic emergency seat (original codename 'rose garden'). Although it was deemed unsafe by engineers on a number of occasions (Bauamt Bonn der Bundesbaudirektion Berlin 1961; Bundesinnenministerium 1969), the project was not abandoned.[6]

It is important to note that there are a number of dark connections between the nuclear bunker and the concentration camp other than the shared location. Firstly, the bunker at Marienthal was run by a number of men who had a track record in an earlier era's total war. Theodor Busse, commander of the Wehrmacht's 9th army and General Walther Wenck, commander of the 12th army, were both involved to the last minute in Hitler's self-destructive Battle for Berlin. The bunker's chief of security was a man called Theo Savaecke, who had been an SS Hauptsturmführer in the war. Convicted in 1999 for war crimes, he was also known as the 'butcher of Milano' (Breitman et al. 2005: 169). Secondly, the bunker's construction consortium comprised a number of companies that had been involved in the erection of the Third Reich's network of camps and bunkers (Bundesinnenministerium 1965). Holzmann AG, for instance, was involved both in the construction of the Atlantic Wall as well as in that of the Monowitz slave labour camp, also known as Auschwitz III (Pohl 1999: 261). Hochtief AG, another member of the consortium, used slave labour for a whole range of major projects during Hitler's twelve-year Reich, which included not just the *Westwall* of defensive fortifications and tank traps, but also the two most iconic of the Third Reich's bunkers, the *Wolfsschanze* headquarters in Rastenburg (now Ketrzyn, Poland) and Hitler's tomb, the *Führerbunker* in Berlin (Hochtief 2014).[7] The most controversial member of the consortium, as noted by Diester (2009, 130), was the Huta AG, also known to have built the crematoria at Auschwitz-Birkenau. Although they hold little explanatory power, the above continuities hint unsubtly at the intimate connection between the camp and the bunker. Whilst these two exceptional spaces emerged in the same place, they also stood at the same nexus of bio- and geopolitics. The following section will unpack these two spaces in more detail to reveal how the nuclear bunker both reproduced and folded the camp inside out.

## Overlaps and Inversions

Like a concentration camp, the nuclear bunker at Marienthal was a tightly sealed and hygienic space that existed both within a network of modern logistics and a landscape of total war. It pre-emptively materialised Agamben's state of exception, thus blurring the boundary between democracy and authoritarianism. Yet, the nuclear bunker was also characterised by a different temporality to the camp, another *telos*, a distinct 'final solution', and it is here that we can begin to understand the way in which the nuclear bunker stretches biopolitics to its vanishing point.

Most importantly, the bunker was subject to high concentrations of sovereign power, crystallising in an exceptional geography and the exposure of life to death. The latter is particularly visible in a string of NATO exercises that were performed in the bunker between 1966 and 1989 and to which we will return in more detail in Chapter 6. All of these simulations operated with similar storylines, typically starting with instability in the Eastern bloc, after which this geopolitical volatility would lead to superpower tensions and a deployment of Soviet troops at the German–German border, followed by a Soviet attack. After a period of conventional warfare and the use of chemical weapons by the Soviet Union, NATO would respond with nuclear strikes in an attempt to force the Red Army to retreat. These exercises simulated the transition into a state of emergency and the decision to deploy nuclear weapons and sometimes the entry into all-out nuclear war. In theory, West Germany and other member states had the right to consult the United States in its sovereign decision over the use of nuclear weapons. Given, however, that Germany itself was to serve as the battlefield of this war, this privilege only highlighted its lack of sovereign control over its own territory and population.

Nevertheless, even during the 1980s, simulations still necessitated the West German placebo sovereign to use military violence against its own population, especially against peace protestors and strikers (Der Spiegel 1985a: 15).

Whilst geographers have stressed the topological and unlocalisable nature of the state of exception (Belcher et al. 2008), it is important to note that there are spaces like the nuclear bunker in which the state of exception materialises quite unambiguously. In the West German case this was through the enactment of the controversial 1968 Emergency Acts (*Notstandsgesetzgebung*), which ensured the government's ability to act in an emergency such as a nuclear war. These laws, briefly mentioned by Agamben (2005: 11), allowed the young West German liberal democracy to strengthen its executive and effectively morph into a semi-authoritarian state during a state of exception, similar to the Bush administration's emergency decrees after 9/11. To be precise, however, the bunker was not so much a space that emerged through emergency decrees, like Guantanamo Bay, than a space in which these decrees could be playfully enacted even before they were passed in parliament (we will return to this in what follows). Like the camp, the bunker operated in limbo between secrecy and transparency, securitised exception and liberal norm. Whilst it was hidden from the public eye and obscured within the federal budget (Bundesinnenministerium 1973), information about the bunker was frequently leaked to the press. Although it was not strictly speaking an extra-legal space, the question of ownership was at least ambiguous; the national rail company still owned the tunnels whilst winemakers still owned the slopes above the bunker. Interestingly, the ministry dismissed the problem of ownership by referring to how the World War II labour camp in Marienthal had been run under similar conditions. In a 1960 letter to the Ministry of Transport, the Ministry of the Interior ordered that the matter should be pushed forward 'without any bureaucratic inhibitions' (Bundesinnenministerium 1960c).

Like the subterranean concentration camp, the nuclear bunker at Marienthal was a site of advanced logistics, both in its peacetime organisation and in the planning for life during nuclear war. It protected its occupants from a range of threats, including a nuclear blast wave, radioactive fallout, as well as biological and chemical agents. In order to function as a tightly sealed and autarkic space, the structure relied on a whole range of technologies, such as blast doors, reinforced concrete, CO alarm systems, radiometers, ABC filters and decontamination facilities. At the same time, the bunker had to provide the supply of oxygen, food, and power as well as discharging waste in order to make human survival possible inside this hermetically sealed capsule. In order to increase its resilience, the bunker was divided into two segments, both of which were to be 'viable' (*lebensfähig*) (Deutsche Societät Beratender Ingenieure 1961). A highly sophisticated filter system alongside a spatiality of quarantine ensured that radioactive dust, pathogens or chemical agents would not penetrate the bunker's interior. The filter room was only to be entered with a protective suit and with explicit permission from the bunker commander (ibid.). Furthermore, the bunker included offices, hospitals, a dentist, a church, a cinema and even a hairdresser to maintain the physical and spiritual well-being of its occupants. Life in the bunker was planned to the last detail, from meetings and meals to hairdresser appointments. Water was rationed and the amount of personal and office space was meticulously calculated depending on rank. Entry into the different parts of the building was tightly regulated by different identification cards (Der Spiegel 1984: 73).

This logistic apparatus, which aimed to ensure a hermetically sealed space, resembled that enabling the camp (Figure 4.6). As Giaccara and Minca (2011a: 5) have argued:

> The rigid separation between the camp and its exterior was paralleled by an obsessive calculative management of the interior. The internal spatialities of the camp – from the dormitories to the latrines – were planned in detail in order to minimize the consumption and the use of space, and to maximize control and discipline (Giaccaria & Minca 2011a: 5).

In this way, ministerial files from the 1960s and 1970s reveal little tables with human labour and partitioned space, technical jargon, long lists of resources and components as well as complaints about supply shortages and a general lack of

**Figure 4.6**    Blast door. Source: Author, 2013.

secrecy around the site. What is particularly striking about these calculations is an obsession with cleanliness. Indeed, even the cleaning of offices was carefully calculated (Bundesbaudirektion 1968a). The question of hygiene also surfaced in relation to the toilets and the sewage system. Already in 1959 ministerial records show that local communities had suffered from unpleasant odours when sewage from what was then the camp was discharged into the local river during World War II (Bundesinnenministerium 1959b). The issue emerged again during the site's construction and again in 1967, a year after the government had paid an unnamed 'institute of hygiene' to test the sewage plant (Bundesinnenministerium 1966). By 1967, the site's sewage had turned into what the local administration called a 'political issue' (Stadtverwaltung Ahrweiler 1967). Interestingly, it was not the bunker's elitist and exclusive logic or the political insensitivity of constructing it on the site of a former concentration camp that caused this outrage but the bunker's lack of hygiene. In response to this pressure, the treasury decided to build what it called a 'federal sewer' (*bundeseigener Abwasserkanal*), an exceptional tunnel that bypassed the local population's sewage system, highlighting the privileged position of the sovereign's bunker (Bundesschatzministerium 1967).

Whereas the nuclear bunker at Marienthal resembled the archetype of the camp as a space of logistics and exception, in other ways it folded the camp inside out. As has been noted above, the camp and the bunker are both logistic and hygienic spaces. In other respects, the two living spaces are hardly congruous. In fact, the nuclear bunker in Marienthal directly contradicted Friedrich Ratzel's influential formulation of *Lebensraum* that had stressed the importance of open space. In a famous 1901 essay he claimed that 'vast spaces are life-sustaining' (Ratzel 1901: 169) but warned that the 'struggle for life' [*Kampf ums Dasein*] became 'desperate in narrow spaces' (ibid.: 153). The dissonance between fascist and Cold War geopolitics is most clearly understood as a spatial inversion of life and death. The nuclear bunker was devised to protect its concentrated living space inside from the holocaust outside, whilst Nazi living space was racially 'purified' through the genocidal logic of the camp. This inversion is especially visible in Marienthal's decontamination facilities and its watchtowers.

As the building authorities ordered, the bunker's main exits had to include decontamination facilities (Figure 4.7), which could be bypassed in times of peace, but which had to be passed in a state of emergency (*Katastrophenfall*). Guidelines detailed a procedure whereby a contaminated person's hands and feet were examined by a radiometer after which their clothes were thrown into a chute and the body was showered in citric acid (Bauamt Bonn der Bundesbaudirektion Berlin 1961). The Ministry of the Interior cautioned in unmistakably biopolitical tones that individuals would need to be subjected to sophisticated tests in order to see whether radioactive materials had entered beyond the clothing and into the body. The document emphasised that this

**Figure 4.7**    Decontamination showers at Marienthal. Source: Ausweichsitz der Verfassungsorgane der Bundesrepublik Deutschland, Dirk Vorderstraße, 2019, https://commons.wikimedia.org/wiki/File:Ausweichsitz_der_Verfassungsorgane_der_Bundesrepublik_Deutschland_(10563686356).jpg. Licensed under CC-BY-2.0

was 'important both for the person affected and for the surroundings' (Bundesinnenministerium 1961). This would suggest that such a highly contaminated person might not be permitted to re-enter the bunker and would thereby be abandoned, reducing that person to 'bare life', following Agamben's formulation. It is perhaps hardly coincidental that these decontamination showers functioned to invert those in Nazi extermination camps. As at Auschwitz, these showers were integrated into a biopolitical machine that medicalised the human body and claimed to 'decontaminate' or 'delouse' it. It is here that a sovereign decision would be taken on whether an individual would be let inside the bunker or left outside to die. Interestingly, in 1968 scientists ran a number of tests with the bunker's filter system. One of the test substances was hydrogen cyanide, also used in the gas chambers under the name 'Cyclone B' (Bundesbaudirektion 1968c).

Although officially run by the *Bundesamt für zivilen Bevölkerungsschutz* (the Federal Office for Civil Defence), the bunker in Marienthal was in fact only designed to accommodate the bureaucratic elites. Underneath a façade of care for life (*Bevölkerungsschutz* translates literally as 'the protection of the population'), the bunker was in fact a space devoid of its population. NATO exercises that were played in the bunker treated the population as a mere obstacle to the smooth functioning of military operations (Thoß 2007: 45). In January 1968, the Federal Building Office first asked for watchtowers to be built, demanding protection from conventional warfare and noting in particular the threat from missiles and artillery projectiles (Bundesbaudirektion 1968b). This happened at the height of the West German student protest movement and only months before the controversial emergency laws were passed in the West German parliament. It is therefore worth asking whom these fortifications were directed against. After all, the watchtowers would hardly have withheld the advancing Soviet army and would, if at all, have drawn unnecessary attention to the bunker.[8] Given that the NATO simulations required the stamping out of various forms of public unrest, it is more than probable that these fortifications were built to prevent armed citizens' entry into the bunker. Whereas the concentration camp watchtower prevented those on the camp's inside from escaping and 'contaminating' the *Lebensraum*, the watchtowers in Marienthal prevented the German population from gaining access to the protective living space of the bunker.

In this way, then, the bunker revealed itself as a space of pure sovereign power that was devoid of the original object of biopolitics, the population. Although it appeared on paper as a civil defence measure, the population's protection was in fact no more than a fig leaf for the shelter of naked sovereign power, its bureaucratic staff, its typewriters and its filing cabinets. Whilst the bunker was kept secret from the population, the existence of *a* governmental bunker was not. Indeed, the civil defence literature was keen to stress that even if an individual or a larger part of the population were to die, the state would be able to survive underground (no author 1962: 92). In this sense, then, the governmental bunker was an obsessive *Raum ohne Volk* (space without people) that inverted the interwar fixation with its opposite – *Volk ohne Raum* (people without space). Whilst there was a quasi-parliamentary assembly in the bunker (the *Notparlament*), its power had been reduced to that of a placebo sovereign. The West German bunker, therefore, was the place where the *bios* and the *polis* collapsed into one another. If the concentration camp had once functioned as the space where sovereign power was reproduced, then the nuclear bunker would have served as its tomb.

Ironically, in the case of a nuclear war, the very weapon that had once been assembled by slave labourers in the tunnels of Marienthal would have returned to haunt the underground structure. The German rocket scientist Wernher von Braun's technologically radical V2 rocket had of course been a crucial last hope for the Nazi regime towards the end of World War II. The story of von Braun is well documented and his work on both the Nazi and Cold War American rocket programmes made him a symbol for the continuity of Nazi scientific fantasies into the Cold War (MacDonald 2008: 617). What is perhaps less well known is the fact that the V2 also served as a prototype for the first Soviet Intercontinental Ballistic Missile (ICBM). Rather than defend a racially purified living space, this weapon would have returned to transform Marienthal into a wintery landscape of extermination.

Silvia Berger Ziauddin (2016: 3) has argued that the nuclear bunker is a *temporary* space of transition. In operation only temporarily, it is indeed built to guide its occupants into a post-apocalyptic world. And yet, as a societal fantasy, the nuclear bunker materialises like few other places a *permanent* state of exception that is at the heart of the Cold War state and modern biopolitics more generally.

## Opening

Before we move on to a discussion of the spaces of nuclear annihilation, it is important to highlight some specificities of German civil defence vis-à-vis the United States, which remains the focus of most explorations of Cold War architecture today. Unlike in the United States, the West German bunker fantasy could draw on the German population's experience of bunkered life during World War II. Somewhat paradoxically, this meant that the government felt the need both to euphemise the effects of nuclear weapons and to sell the bunker as the viable solution against these effects. Even in the age of the H-bomb, the Ministries of the Interior and the Ministry of Defence would argue, protection was possible and moreover important to combat defeatism (Schröder 1954; Bundesministerium der Verteidigung 1956: 1). In 1957, the West Germans entered into an agreement with the United States, so West Germany too could test how a number of different German bunker models would respond to a nuclear blast in the Nevada desert. The government's final report came to the conclusion that 'protection is possible against the effects of every weapon' (Bundesinnenministerium 1958a). Tests with human subjects would moreover reveal that although bunker occupants would suffer from a wide range of discomforts in the event of a retreat into subterranea, it was generally possible to spend up to five days in the Federal Republic's civilian bunker types without any serious physical or psychological breakdown (no author 1959). The government stance was thus summed up in the title of the pamphlet *Everyone Has a Chance* (Bundesamt für zivilen Bevölkerungsschutz 1961a).

For the technophile theorists of a nuclear Germany, the protection of the population from nuclear weapons would be achievable in the not so distant future. Adenauer's scientist Pascual Jordan who had dreamed of underground cities in the mid-1950s would argue that

> Of course, the explosion of an atomic weapon results in the release of immense energies, which have very unpleasant effects on major cities. But if you compare that to the forceful eruption of a volcano, then even the hydrogen bomb is but of toylike irrelevance (Jordan 1954: 63).

Ultimately, civil defence wanted to offer biopolitical 'methods and facilities' that allowed 'for a certain sense of immunity against nuclear catastrophes' (Jordan 1957: 174). 'Of course, it is very unpleasant that radioactive contamination would lead to small doses of Strontium being ingested, causing bone tumours', Jordan noted, 'but it is only a considerable lack of fantasy that brings us to assume that bone tumours will still be a serious illness in thirty years' time' (ibid.: 174).

And yet, there were always sceptical voices too, especially in the media (Molitor 2011: 78; see also Lemke 2007: 79), and amongst scientists and anti-war activists (Stenck 2008). In 1960, even the West German Ministry of Finance warned that modern warfare made absolute protection through bunkers impossible (Bundesfinanzministerium 1960). The Ministry of the Interior disagreed with this verdict, highlighting again the particular condition under which West German civil defence had to operate.

> Unlike the United States, we have to assume that the Federal Republic will become the theatre of operations in the case of war [...] If we compare the population density of the US with that of the Federal Republic, a crucial difference emerges: The United States [has] approximately 20 inhabitants per square kilometer, the Federal Republic [has] 200 per square kilometer (Bundesinnenministerium 1960: 4).

Ultimately, the idea of civil defence was based on the assumption that modern warfare had collapsed the distinction between combatant and non-combatant. 'The atmosphere of war', one bureaucrat noted, 'surrounded each individual', thereby obstructing the totality of the society's 'vital processes', particularly labour (Mielenz 1953: 5). Civil defence was organised on the paradoxical insistence on the awesomeness of nuclear weapons on the one hand, and on the possibility of nevertheless defending oneself against them, on the other. 'A central project of civil defence', Joe Masco has argued, 'was thus to produce fear but not terror, anxiety but not panic, and to inform the public about nuclear science but not fully educate it about nuclear war' (Masco 2014: 52). The bunker expressed this logic in a way little else could.

> The solidity of the reinforced bunker invokes both the power of the weapons it is intended to repel and something of the folly of the attempt to seek any sort of shelter from them. As such, the bunker is both a sign of industrialized warfare and of there being no escape from it (Beck 2011: 81–2).

We can sense in West Germany's preparation for nuclear war a real urge to read the coming war through the lens of the previous one. During the 1950s, nuclear weapons had often only been mentioned in passing. The issue was still framed through the experience of conventional bombing campaigns during World War II. And so, from the 1950s onwards, the nuclear bunker could re-emerge as a key architectural focal point, a space that promised to protect lives in a conflict that threatened to rid the earth of life itself. In many ways, the nuclear bunker replicated the spatial logic of the camp, that iconic architectural space that stood at the vanishing point of Nazi geo- and biopolitics. Like the camp, the bunker was an Agambenian space of exception within which total war and extermination were rendered rational. Indeed, the governmental nuclear bunker was enabled and governed by a highly concentrated form of sovereign power, by a logistic apparatus that was obsessively committed to the creation of hermetically sealed and hygienic living space. Yet, despite these congruities, the Cold War inverted the fascist relationship between the vast living space and the confined space of the camp, thereby gearing the bunker to a different *telos* to the camp. Its 'final solution' was a tightly sealed sarcophagus for sovereign power, a space without people, a *Raum ohne Volk*.

In this sense, the nuclear bunker needs to be understood as a geo- and biopolitical space in its own right, for it stretches Esposito's logic of immunisation to its vanishing point. In the thermonuclear age, with very few exceptions, the entire population were *homines sacri*. Whilst the US civil defence programme tried to sell the bunker as a quintessentially American space that materialised both the American frontier in subterranea and the wonders of consumer capital – Americans could extend their suburban houses by an extra room – (Masco 2009), the West Germans framed their nuclear bunkers through familiar geo- and biopolitical concepts. It was only after the Social Democrats formed successive governments during the 1970s that West German civil defence managed to emancipate itself ideologically from its National Socialist predecessor (Molitor 2015: 395).

## Endnotes

1   The office was established in 1952 within the Home Office and changed names a number of times.

2   The Bundeswehr, too, was planning a whole range of bunkers for its forces. A 1959 report, signed by the former Wehrmacht general and head of the new German army Adolf Heusinger, also mentions how different sanitation rooms would need to separate out spaces for those injured by atomic, biological and chemical weapons and those who had merely been wounded by conventional weapons (Bundesverteidigungsministerium 1959: 10).

3   *Ziviler Luftschutz* had been published from 1933 to 1945 under the title *Luftschutz und Gasschutz*.

4   The fact that Cold War civil defence was predominantly framed through geopolitics did not of course mean that it was not also framed through other discourses, such as psychology.

5   In 1960, Löfken would argue that bunkers themselves were not sufficient in securing the survival of the state. The government had to put in place vital infrastructure and resources for a life after a nuclear war (Löfken 1960).

6   The bunker was in operation until its eventual decommissioning in 1997, briefly halted only in the anxious aftermath of 9/11. Since 2008 a small part of the bunker has served as a heritage site.

7   Of course, the governmental nuclear bunker at Marienthal in many ways resembles Hitler's *Führerbunker* in Berlin. Yet, it was particularly the advance of nuclear war and its logic of extermination that highlighted its connection to the camp as the thanatopolitical underside of biopolitics.

8   The Red Army Fraktion (Baader Meinhof group) can be ruled out at this point. Although the members were already politically active in 1968, the terrorist cell was founded only two years later.

# Chapter Five
# Spaces of Extermination

*The more life is prepared for death, the more beautiful it becomes*
*Friedrich Ratzel*, 1903

## Places of Forgetting

Memory is an important commodity in post-Cold War Germany. Three decades after its fall, the Berlin wall is one of the world's most heavily memorialised sites, a tourist attraction like few other twentieth-century structures. But whilst the remains of the iron curtain continue to function as *the* symbol of Germany's Cold War, it is nevertheless a highly problematic symbol. Rather than representing the threat of mutual nuclear annihilation, it stands for a much simpler lesson – that of the West's moral victory over the 'prison' of real existing socialism.

Around 200 miles west of Berlin lies the Point Alpha observation point, formerly home to the US 11th and 14th Armoured Cavalry Regiments. The now abandoned military site is situated at the heart of the so-called Fulda Gap, a lowland in the central German section of the iron curtain. During much of the Cold War, NATO strategists were convinced that the Warsaw Pact would strike here on its way to the Atlantic. Today, Point Alpha has been turned into a museum that combines a reconstruction of the East German border fortification with an exhibition of US armoured vehicles. But here, too, the message is surprisingly simple. Whilst the risk of a nuclear war on German soil is only mentioned in

*Cryptic Concrete: A Subterranean Journey Into Cold War Germany*, First Edition. Ian Klinke.
© 2018 John Wiley & Sons Ltd. Published 2018 by John Wiley & Sons Ltd.

passing, visitors learn instead that the threat of 'mutual destruction helps prevent nuclear war'. Perhaps unsurprisingly, a memorial in the centre of the museum thanks the US Army for its service for 'peace and freedom'. One has to travel another 40 miles west of Point Alpha to find a more telling site.

Visitors to SAS Alten-Buseck are greeted by a double barbed wire fence and a watchtower (Figure 5.1). The abandoned military installation has been reclaimed by the forest and much of it is covered by long grass and shrubs. Broken glass cracks as one enters the main area of the camp through the former barracks. Empty beer cans litter the floor and there are signs of vandalism, but there are also images of palm trees on the wall that convey a sense of uncanny homeliness. The end of a dog leash and a few kennels indicate that the camp was once not only inhabited by humans. As one ventures across the inner site, a number of concrete fortifications and earthwork come into vision, a further watchtower and a ring of lanterns that stands just behind the double barbed wire fence. A frog is croaking in a trench that has now turned into a lake. At the bottom of the field there are two larger bunkers, secured with double steel doors. It is behind these doors that the precious nuclear weapons would have been stored (Figure 5.2). As one leaves the site, it is difficult not to agree with Rachel Woodward (2004: 152) that 'these places are emblematic of domination, power and control', even long after they have been abandoned.

**Figure 5.1**    Watchtower at Special Ammunition Site Alten-Buseck 2015. Source: Author.

**Figure 5.2**    Warhead storage bunker at Special Ammunition Site Alten-Buseck 2015. Source: Author.

SAS Alten-Buseck was one of NATO's Special Ammunition Sites, used for the storage of nuclear weapons during the Cold War. Built in the 1970s and abandoned in the 1990s, Alten-Buseck was merely one of a whole range of nuclear weapons camps that continue to litter the Central European landscape. Alten-Buseck hosted nuclear warheads for three weapons systems, the nuclear-capable howitzers M109 and M110 as well as the surface-to-surface missile 'Honest John' (Atomwaffen a-z 2015). Although technically 'just' a tactical nuclear weapon designed for battlefield use, the latter could have yields of up to 100 kilotons, almost seven times the destructive power of the bomb dropped on Hiroshima in 1945. Perhaps surprisingly, the missile had a reach of only around 30 miles. Tactical nuclear weapons stored in West Germany would thus have been able to easily wipe out large parts of – if not the entire – German population. And yet, sites like Alten-Buseck do not attract much attention in contemporary Germany. A nation that has done so much to re-examine its history of mass annihilation during World War II seems to have tried to erase its memory of these deadly spaces of extermination.

This chapter journeys into NATO's now abandoned Special Ammunition Sites, used from the 1950s onwards to store short-range tactical nuclear weapons in Europe. It starts by tracing the relationship between the Federal Republic and the atom bomb, zooming in on NATO's nuclear sharing initiative. Following this, the chapter moves on to a dissection of the politico-material space that emerged as a direct consequence of West Germany's participation in NATO's nuclear sharing: the tactical missile storage camp. Through an examination of a number of sites, I argue that these spaces of exception articulated not just the geopolitics of the Cold War but a biopolitical concern with national survival and extermination, even self-annihilation. The short ranges of the weapon systems stored in these highly securitised camps and the fact that they were situated so far inward from the iron curtain meant that these missiles could only really be used on the Federal Republic's own territory – with detrimental effects for its own population. The chapter concludes that as a space of (potential) extermination, the tactical nuclear missile camp managed to solve one of the fundamental logistic problems of the Nazi death camp – the production and disposal of the corpse. Nevertheless, it was the impossibility of concealing its barbed wire and watchtowers in the landscape that led to it becoming increasingly targeted by the emerging peace movement.

Much as today, the stationing of nuclear weapons in West Germany was legitimated by a doctrine of deterrence. But it is important to remember that nuclear weapons were initially built for very different reasons, mainly to demonstrate power and to submit an enemy into giving up. As E. P. Thompson reminds us in his deconstruction of deterrence theory, deterrence only became a dominant discourse *after* the Soviet Union had acquired thermonuclear weapons in 1953. Rather than an attempt to intellectualise security in the nuclear age, he claimed, it was more of an excuse to expand American stockpiles. To speak of a nuclear weapon as a 'deterrent', was thus to 'ascribe the intentions (or purported intentions) of the users to the weapon itself, as part of its inherent quality' (Thompson 1982b: 6). So, if deterrence cannot be the answer to the question of why West Germany was so heavily nuclearised, then we need a different theory. This chapter proposes that at the heart of the tactical nuclear weapons stationed in West Germany was an impulse towards biopolitics and its logic of survival and extermination. The very obsession with life and death that had been turned into a geopolitical theory by German geographers in the late nineteenth and early twentieth century and that had materialised in the architectural spaces of the bunker and the camp during the early 1940s was emerging in a new form in the 1950s. The second attempt to materialise a politics of national survival and extinction had arrived.

## Sharing the Bomb

In her history of the US intercontinental missile 'Minuteman', Gretchen Heefner (2012) shows how most American citizens who lived near to nuclear missile silos during the Cold War had no problem with the missiles and often even endorsed their presence as a patriotic act. Today, many front gardens of villages that lie near abandoned nuclear missile sites in West Germany greet strangers with the national flag of the United States of America. This open identification with the power that nuclearised West Germany is perhaps puzzling if one considers that these sites would have been primary targets for the Soviet Union, had the Cold War turned hot. And yet, it is also telling about the relationship between the West German state and its transatlantic ally.

Whilst West Germany had initially been keen to develop its own nuclear weapons programme, NATO had forced West Germany to renounce any ambitions of acquiring atomic, biological or chemical weapons and had also denied Bonn the production of delivery systems. Konrad Adenauer announced in 1954 that the Federal Republic would voluntarily renounce the manufacture of atomic, biological and chemical weapons (Bulletin 1954). But the West German leadership nevertheless wanted some form of nuclear co-ownership, both in order to gain more influence on US decision-making and to achieve a higher status within the alliance that would conceal the absence of nuclear muscle (Hoppe 1992: 371).

By the early 1950s, the United States was starting to nuclearise West Germany. Following its policy of *Westbindung* and its enthusiasm for nuclear weapons, the West German government had few objections. In 1953, the West German weekly TV programme *Welt im Bild* enthusiastically celebrated the arrival of the first six M65 atomic cannons, which had recently been tested in a Nevada site. 'Atomic Annie' could deliver nuclear warheads with a yield that approached that of the Hiroshima bomb and was praised for its 'wonderful adaptability to the local context' (Welt im Bild 1953). This local context was of course Germany itself. Only in passing does the programme mention that the nuclear cannon had a reach of no more than 22 miles. In order to build a consensus for West Germany's nuclearisation, the state euphemised these new weapons.[1] In 1956, the Federal Ministry of Defence wrote in an internal document that tactical atomic warfare 'did not invalidate established principles of military tactics' but only altered behaviour in the battlefield' (Bundesministerium der Verteidigung 1956: 1). A year later, Adenauer would famously declare that tactical nuclear weapons like Atomic Annie were 'nothing but the further development of the artillery' (Der Spiegel 1957a).

After the strategy of Massive Retaliation was introduced in 1957, even conventional attack would have been countered with tactical nuclear weapons. In the same year, the North Atlantic Council established its nuclear sharing initiative, which continues to govern the nuclear cooperation between NATO allies to this day. It allowed non-nuclear powers like Germany, Turkey, Belgium, the Netherlands and Italy to participate in nuclear planning, stationing and delivery of nuclear weapons. It was difficult for the Allies to imagine a nuclear-armed German state given how the Third Reich had famously worked on the atom bomb but ultimately fallen behind the US and British nuclear programmes. And yet, the idea of nuclear sharing seemed to be a compromise. Bonn's allies did not have to fear a German bomb and West Germany could at least simulate some form of nuclear status. The Supreme Headquarters Allied Powers Europe (SHAPE), after all, was clear in that the weapons now distributed throughout Europe were 'designed to support NATO defence planning rather than national planning' (SHAPE 1974: 3).[2] The Bundeswehr would come to frame this participation through the oxymoron 'offensive defence'. As General Heusinger would explain in 1957, the fact that the Bundeswehr was now equipped with nuclear capable artillery had led to a blurring between attack and defence (Der Spiegel 1957b). The army saw its soldiers no longer simply as the 'shield' that was to accompany Washington's nuclear 'sword'.

And yet, whilst Bonn's elites were eager to participate in the Western world's new nuclear geography, the German population was much less enthused. After Adenauer's conservative government passed a law in 1958 that allowed the Bundeswehr to own carrier systems for nuclear weapons, the population took to the streets. The personal memories of the annihilation of German cities by Allied bombing campaigns only a decade earlier were simply too fresh and soon the parliamentary opposition pushed for a referendum on the question. This, however, was rejected by the Federal Constitutional Court. The Federal Republic was determined to participate in NATO's new nuclear geopolitics. The Ministry of Defence noted complacently that the handover of foreign nuclear weapons to the German armed forces had been declared 'legal, both provisionally and as a final provision' (Bundesministerium der Verteidigung 1958: 2). In those years, Germany witnessed the birth of its anti-nuclear movement, which would come to the fore especially in the early 1980s.

The ensuing nuclearisation of the Federal Republic did not go unnoticed on the other side of the iron curtain. By 1959, the East Germans noted that there were a whole range of nuclear weapons systems on West German territory, including not just the aforementioned surface-to-surface tactical nuclear missile ('Honest John'), but also the United States' first guided nuclear ballistic missile ('Corporal'), its first large tactical missile ('Redstone') and the nuclear-capable cruise missile ('Matador'), as well as a range of different Howitzers (Ministerium für Nationale Verteidigung 1959: 5–6). By 1965, NATO had placed 500–550 such nuclear land mines on the West German side of the iron curtain and a further 200–240 nuclear weapons further inland. As it has emerged, it was the Federal Republic rather than the United States that actively sought to install this cordon of so-called 'atomic demolition munition' (Bald 2008, 52).

West German opposition to these developments first emerged in the 1950s when scientists like Max Born and Werner Heisenberg prompted the Adenauer government to launch a PR campaign to counter the growing public unease with nuclear weapons. Elsewhere physicists, who had since the late 1940s congregated around the *Bulletin of Atomic Scientists* and its powerful chronopolitical device the doomsday clock, raised similar issues. After resistance against nuclear weapons had become more organised in the United Kingdom around the 1958 Campaign for Nuclear Disarmament (CND) (see Featherstone 2012; Herb 2005), a broader protest movement emerged in West Germany, too. In the 1960s, this *Friedensbewegung* ('peace movement') united communists and Christians, artists and scientists, disillusioned generals and dissatisfied civilians.

Despite these protests, which fed into the attempted 1968 student revolution, successive West German governments stuck to a policy of hard-line deterrence coupled with the controversial non-recognition of the GDR (the so-called 'Hallstein doctrine'). In the 1970s, Social Democrat-led governments supplemented Bonn's traditional policy of *Westbindung* with one of *Ostpolitik* ('Eastern policy'), which sought to relax and normalise relations with the East bloc. The question of West Germany's role within NATO's nuclear posture was largely kept quiet during this time and was only to arise again under the Social Democrat-led government of Chancellor Helmut Schmidt (1974–1982). Schmidt had played an important role in abandoning the atomic demolition munition in the late 1960s (Bald 2008), but in the early 1980s became the champion of a new nuclear arms buildup. Schmidt saw the Federal Republic as threatened by the Soviet Union's new SS20 weapons system and was hoping that new intermediate range nuclear weapons (Pershing II) in Central Europe would force Moscow to move their missiles further inland. The talks under the new Reagan administration failed and the new weapons, which would leave Moscow with near to no time to react to an attack, were stationed on West German territory in the mid-1980s.

After the phase of 1970s *Ostpolitik*, the 1980s brought a second wave of nuclearisation in Germany as well as a second wave of anti-nuclear protests, which were sparked by NATO's new Pershing missiles. In the course of these protests, the Friedensbewegung staged so-called 'die-ins' – highly visible protests in public squares and in front of military installations – and connected cities with 'human chains'. It is important to note the wider public support for this anti-nuclear struggle. Between 1980 and 1983 around 4 million West German citizens signed the so-called Krefelder Appell, which called on the Federal Republic to repeal its support for the stationing of intermediate range missiles in Western Europe. In October 1983, *Der Spiegel* found that around two-thirds of the West German population were opposed the nuclear arms race (Der Spiegel 1983). It is in this context that a highly destructive space that had first been established decades earlier – the tactical nuclear missile camp – came under increasing public scrutiny.

## The Architecture of Missile Storage

During the Cold War, West Germany became one of the most heavily nuclearised territories in the world. Anti-war activists calculated in the mid-1980s that there were around 3,400 nuclear warheads in the Federal Republic and 241 installations that were either used to store nuclear weapons or to support their stationing (Arkin & Fieldhouse 1986: 275). The list of weapons systems that were stationed in West Germany was long and included not just US but also British nuclear weapons and involved Belgian, Dutch and Canadian forces, too. The key places through which NATO's nuclear sharing initiative operated were the Special Ammunition Sites, medium-sized stationary US Army and NATO installations for the storage of nuclear warheads, missiles, gravity bombs and demolition munition. In a transition from peace to a state of war, nuclear ammunition would have been removed from these fixed storage facilities and dispersed to mobile weapons holding areas or nuclear ammunitions supply points. Whilst camps were made to 'blend into' the countryside, their architectural design would emerge in a very recognisable spatial grid that conformed to the guidelines provided by NATO. Given the highly securitised nature of the camps and the on-site cooperation between US and West German forces, the camps were subject to a complex overlapping of sovereignties and a legally exceptional geography.

Despite the fact that the locations where nuclear weapons were kept were technically secret, the militarisation of the landscape did not go unnoticed. Both the peace activists and the Stasi noted how easy it was to distinguish nuclear weapons camps from structures that merely held conventional weapons (Informationsbüro für Friedenspolitik 1982: 10; Ministerium für Nationale Verteidigung 1972: 4; see also Schregel 2011: 91). In the 1980s, when the debate around Pershing II emerged, the anti-nuclear movement started to take a more active interest in these sites, producing guides on where to locate existing nuclear weapons camps and how to detect new ones (Luber 1982). The East German secret service, too, had obtained detailed descriptions of the camps; plans and images of sites as well as maps that showed their location. Nuclear missile camps generally fell into three classes with similar infrastructure and commando structures. The first were the custodial storage sites which supported delivery units (SAS sites types A, B, C, F, and UK SSA). These would perform weapons assembly, warhead mating (sic), and limited maintenance. Second were the support and depot sites, which performed maintenance and stored reserve stocks of weapons and components. Thirdly, supply depots (which held the actual warheads), were located between 50 and 140 miles from the iron curtain so as to ensure enough time to deploy the missiles in case of an attack.

Facilities for the storage, maintenance, and custody of nuclear weapons were so similar because they were planned and built by 'a common framework' for siting, planning, design and construction. A survey would first 'examine the terrain proposed for the sites, the surrounding countryside, and the general locale', after which sites were selected that were 'capable of accommodating the facilities prescribed by criteria and depicted on drawings without creating unfavourable security, custodial, or operational problems' (SHAPE 1974: J-2). Whilst the exact layout of any site would depend on 'the terrain, size of the area and the mission', all sites had to be situated within a capable road network, be defensible, and

**Figure 5.3**    NATO atomic missiles storage (1992). Source: Author.

isolated from builtup areas and troop concentrations (US Army 1984: 33, 39). Many of these sites thus ended up being hidden in forests (see Figures 5.3 and 5.4).[3] As the East German secret service found out, local conditions were used to camouflage nuclear weapons sites and, as a general rule, the vegetation had to stay as untouched as possible. Depots were meant to be located in steep valleys and canyons and always secured by semi-sunk concrete bunkers (also known as 'igloos') that were covered with earth (Ministerium für Nationale Verteidigung 1972: 9).

Everything was organised according to a spatial grid. The bunkers would be ordered symmetrically, either in a circle or a square. Whilst the number of igloos could vary, their spatial design and the distance between them was pre-given and only transformed slightly over the cause of the Cold War. The bunkers measured 18.5m × 8.3m × 4.0m and had doors measuring 3.05m × 3.03m so a truck could enter. Minimum spacing requirements were designed 'to avoid excessive interaction of radioactive materials in adjacent arrays of weapons components' (US Army 1984: 32). Although nuclear missile camps were not protected against a direct nuclear hit, they were often protected against some of the effects of nearby nuclear warfare (Ministerium für Nationale Verteidigung 1972: 4).

**Figure 5.4**    Guards at Camp Bellersdorf (early 1980s). Source: Author.

During peacetime, the West German host nation tried to make the soldiers' lives as 'normal' as possible – from the provision of basic accommodation to schooling (Bundesministerium der Verteidigung 1961a: 37). Life was strictly regimented and standardised, down to the nature of the bins, fridges, bed linen and curtain rails (Bundesministerium der Verteidigung 1962). This was hardly a 'space of loose ends and missing links' (Massey 2005: 12).

Security on site was tight and targeted at a range of wartime and peacetime threats, including aircraft, missile, ground or paramilitary attack as well as espionage, sabotage and civil riots (SHAPE 1974: B-1). In order to counter these different threats, sites were to be protected both with security guards and combat units. Whilst the bunkers were hardened against missile attacks, intruders were warded off by fences, concertina wire, entrance control points, artificial lighting, intruder detection devices, dogs and defensive works. Each camp was to be secured by at least one rifle company (around 150 men) and supplied by the host nation, in this case West Germany (Bundesministerium der Verteidigung 1961a: 15). In peace times, there were augmentation forces, a sabotage alert team, standby alert forces and backup alert forces (US Army 1964). Moreover, counter-intelligence support would continuously survey the local population, as well as observing bars and recreational facilities frequented by the personnel. An important psychological dimension was the 'intensive security indoctrination' of guards (SHAPE 1974: B-1). The transport of nuclear weapons was tightly regulated, too. Convoys would include

> a radio-equipped vehicle with an armed driver and one guard ahead of the column, at least two armed personnel, one of which [would] be a US custodian, for each vehicle transporting atomic warhead sections, a sabotage alert team of at least five personnel in a suitable radio-equipped vehicle in the middle of the column and a radio-equipped vehicle with an armed driver and one guard at the rear of the column (Headquarters Northern Army Group 1961: 4).

Whilst SHAPE (1974: A4-5) thought of the possibility of a nuclear weapon accident as 'practically non-existent', it did demand 'fully trained and equipped disaster control teams', which had the capacity of being rapidly dispatched to the scene of such an accident.

At the heart of the nuclear weapons storage camp was a complex system of overlapping sovereignties, which derived directly from the transnational nature of the nuclear sharing initiative. As the Supreme Headquarters Allied Powers in Europe (SHAPE) clarified, US nukes remained in US custody until released for use by the US Commander in Chief Europe (USCINEUR) (SHAPE 1974: C-1). Whilst the Special Ammunition Site programme was coordinated by the Supreme Allied Commander Europe (SACEUR), which took care of the actual nuclear strike plans, the user nation was responsible for training and equipping nuclear delivery forces and, in some cases, for the actual delivery, too. Custody here included 'basic responsibility for security of the weapons' and the 'external physical security of those areas where nuclear weapons [were] deployed'. This led to a complicated distribution of responsibilities. West Germany complained that the different treaties that governed the operation and supply of the SAS camps often contradicted one another, which led to 'confusion, continuous inquiries and significant extra bureaucratic work' (Bundesministerium der Verteidigung 1963). It was in particular unclear as to whether the host nation would have to cover the rent, medical and dental care as well as expendables of US personnel. Whilst the host nation promised to 'maintain all furniture and furnishings in an attractive, usable condition' it was up to the US tenants to pay 'for deliberate and grossly negligent damages or damage in excess of fair wear and tear' (Bundesministerium der Verteidigung 1961b: 3).

The US government was liable for particular damage, but certainly not for all. Interestingly, third party claims arising 'from nuclear explosions and radiation hazards' were explicitly excluded from the arrangement (Bundesministerium der Verteidigung 1961b: 22). In 1961, the West German Ministry of Defence tried to produce clarity by stating that:

> The presence of US Army personnel at the SAS Support Site will not alter the command responsibility and authority of the non-US station commanders; but with

respect to US Army units, personnel, equipment and material, all functions of command, control, administration, training and tactical operations will be the sole responsibility of the US Army Commander' (ibid.: 3).

Given the confusing nature of this overlapping of sovereign spaces, the West German government thought it necessary to provide some guidance. One manual was even compelled to define the term *Germany* as 'territory that is subject to the laws of the Federal Republic of Germany' in their annex of technical terms (Bundesministerium der Verteidigung 1961a: 12). It was simply unclear where West Germany began and where it ended.

This system of overlapping sovereignties was reflected in the camp's architecture. Inside the camp, there was an exclusion area (also referred as the 'custodial area', 'no lone zone' or 'X-area') in which the nuclear weapons were stored and which was subject to a 'two-man rule'. As the US Army specified,

This X-area is used to store the nuclear weapons and is to be secured by members of the attached security force. The perimeter of the X-area will normally be designated by the commander. The ECP [entry control point] will be manned by two guards, both of whom must be in the personnel reliability program (US Army 1974: 40).

The aim of this exercise was to ensure that each of the men was 'capable of detecting incorrect or unauthorised procedures' – but also to delineate a space to which only American soldiers had access. To mark out its privileged position within this already exclusionary space, the X-area was enclosed by an inner fence, which was controlled by the SAS site commander. The area in between the two fences was variously referred to as the 'restricted' or 'limited' area, access to which was controlled by the commander and the head of the security guards (Bundesministerium der Verteidigung 1961a: 12).

Around this most sacred space was the 'administrative area', designed for the principle purpose of providing administrative control and safety as well as a buffer area of security restrictions (US Army 1974: 41). This space would typically feature a landing area for receipt of nuclear weapons by air convoy. Located within a reasonable distance from the X-area, this landing zone could be either manmade by the use of equipment (such as bulldozers, chainsaws or axes) or could be an already existing open field. Other areas included vehicle and convoy holding and assembly areas as well as traffic control points, made to screen the entry into the area and ensure the flow of traffic. Around all of this was a 'security area' outside the parameter with military police patrols, listening and observation posts and early warning devices.

This architectural layout led to concerns, particularly after the left-wing terror group Red Army Fraction (RAF) had emerged in the 1970s. In 1975, the Ministry of Defence feared that terrorist commandos had already acquired the technical knowledge to detonate an atomic weapon and voiced their concern to Washington that the German security personnel might be held hostage by the terrorists in the X-area in an attempt to gain control over NATO missiles (Bundesministerium der Verteidigung 1975a: 1). Seemingly concerned with the lives of the hostages, the ministry stated in a follow up internal document that although 'any measures taken by German guards would have to be in line with German law' it might be possible to override national law by enforcing an 'extra-legal state of exception' (Bundesministerium der Verteidigung 1975b: 2). Bonn clearly wanted the right to impose a state of exception on the X-area in the case of an emergency. The Americans, however, were not happy with such an arrangement. Whilst showing some understanding for the German concerns, the US clarified:

> Should hostages be used in an attempt to gain access to, as a cover for the removal of, or to thwart recovery of nuclear weapons, the welfare and safety of the hostages will be considered in determining the actions to be taken. However, the presence of hostages shall not deter the taking of prompt and effective actions to deny unauthorised access, prevent removal, or regain custody of nuclear weapons. (USCINCEUR 1975).

At the very heart of the tactical missile storage camp was thus *not* the protection of the West German population but the 'survivability of nuclear ammunition', as a US Army manual for staff working on these sites specified (US Army 1984: 42). Joe Masco has written about the way in which nuclear militarism makes the bomb comprehensible through analogy to the human body. Nuclear weapons thus 'have "birth defects", require "care and feeding", "get sick" and "go to the hospital", get regular "checkups", "retire", and have "autopsies" (Masco 2006: 80). Such medical language could barely hide the fact that, much like the nuclear bunker, the political space of the Special Ammunition Site was in fact a *Raum ohne Volk* (space without people) that inverted the Nazi preoccupation with *Volk ohne Raum* (people without space). The missile camp's prime purpose was simply to ensure 'a smooth transition from peace to war' (US Army 1984: 4). Nuclear geopolitics was always 'more attentive to the survival of nuclear weapons than to the survival of human beings' (Cohn 1987: 715).

## *Raum Ohne Volk*

In grappling with the tactical nuclear missile camp as a political space, it is important to recognise that nuclear war is always already geo- and biopolitical because it takes into consideration the material properties of both the earth and the human body. Whether a missile is detonated 500 metres above ground or at sea level, whether it is exploded above a wet or dry terrain, produces markedly different results on the built environment and on human flesh. These effects were taken into consideration at different scales of military planning and so the US Army warned its staff who were working on tactical nuclear weapons sites in Germany that casualties would be high and that their units could be 'encircled, cut off, and sometimes totally destroyed'. 'In order to survive', the army explained, each soldier must understand their job and why they were doing it (US Army 1984: 17). Soldiers were not just seen as part of the chain of command but were also recognised as biopolitical bodies. Like the Americans, the West German army was concerned that soldiers were ill-prepared for the effects of nuclear radiation on the battlefield. Thus, the Federal Ministry of Defence debated whether soldiers should be exposed to mild chemical agents during training exercises that might induce the kind of sickness caused by nuclear weapons (Bundesministerium der Verteidigung 1959: 3).[4] We are reminded here of the medicalising logic of biopolitics, which attempts to immunise life through the production of its opposite. The space of the nuclear missile camp was also biopolitical in embodying three other principles: legal exceptionality, exterminism and self-destruction.

Firstly, and as we have already seen above, the nuclear missile camp was a space in which sovereign power and its tendency to reproduce itself through the exception was materialised in an almost pure form. Moreover, the camp's architecture materialised the complex geopolitical relationship between the US government and the West German host nation that emerged from this excess of sovereign power. West German sovereignty ended after all at the margins of the X-area – even though it was precisely here that West Germany's status as a nuclear sharer was most palpable. Yet, it was not just the particular spatial design of the sites that spoke of its nature as a space of the exception. Of course, levels of securitisation around the camp surpassed those of other military sites, but it was the very existence of such a space that highlighted the authoritarian nature of the state of emergency. A nuclear war on German soil would have effectively rendered the Federal Republic subject to military (NATO) command.

It is interesting to see that despite these implications, the West German state was in no way inclined to frame tactical nuclear warfare as illegal. Officer cadets in the Bundeswehr were taught that the use of nuclear weapons was compatible with the law of armed conflict. Given that genocide was seen to be 'the partial or complete destruction of a nation, ethnicity, race or religious group' and 'even large atomic weapons' were 'usually' not seen to be able to do so, the use of nuclear weapons did not fall under the genocide convention (Bundesministerium der Verteidigung 1981: 6). Moreover, West German soldiers would have learned that although the Federal Republic was not permitted to build nuclear weapons on its own territory, this prohibition did not entail 'the use' of nuclear weapons (ibid.: 6). Interestingly, the Bundeswehr made a distinction between, on the one hand, strategic nuclear weapons that targeted civilian populations and which were in violation of international law, and, on the other, tactical 'battlefield' weapons, which were not (35).

In reality, however, the distinction between tactical and strategic nuclear weapons was a thorny issue. In principle, tactical weapons are those with smaller yields and a smaller range that are designed for use on the battlefield whilst strategic weapons are those that are used to take out cities or larger military installations. In practice, however, delivery systems for tactical nukes can have yields that can easily destroy a larger city. The short- and medium-range ballistic missiles deployed in Germany were so crucial because they were mobile and thus less exposed to an enemy missile attack than strategic missiles launched from fixed silos. Somewhat predictably, the Ministry of Defence said nothing of the capacities of tactical nuclear warfare to spiral into strategic warfare – nor of the effects of tactical nuclear warfare on the local population of Central Europe.

Secondly, and perhaps more importantly, the tactical nuclear missile camp was an important aspect of what the historian and anti-nuclear activist E. P. Thompson referred to as 'exterminism'. With this term he wanted to capture the characteristics within a nuclear society's economy, polity and ideology that directed it towards annihilation. The outcome of this social order, he argued, was not accidental (even though the final episode may well have been triggered by an accident) 'but [as] the direct consequence of prior acts of policy, of the accumulation and perfection of the means of extermination, and of the structuring of whole societies so that these [were] directed towards that end' (Thompson 1982a: 20). Everything in the missile camp was geared towards smooth annihilation. Indeed, the dual purpose of these sites was to guarantee the survivability of the nuclear weapons in peacetime and their efficient deployment in times of war. In this, the missile site functioned as a biopolitical space that extended the principles of survival and extermination, elaborated by Ratzel and his followers in the late nineteenth and early twentieth century, to the weapon itself. It is thus no coincidence that the tactical nuclear missile camp resembled the architecture of that other twentieth-century space in which necropolitics had so unmistakably materialised (see Figure 5.3).

The Jewish-American social critic and anti-war activist Marcus Raskin argued in 1982 that nuclear weapons were ultimately 'nothing less than an instant Auschwitz, wrapped in bomb casings and held together by the protons and electrons of the martial and triumphal spirit' (Raskin 1982: 209).[5] Of course, his words were meant as a provocation and an attempt to awaken resistance against a new arms race in the 1980s. And yet, if we look closely at the tactical nuclear missile camp, it is difficult not to spot traces of Auschwitz in this spatial killing machine. After all, nuclear war worked through what Markusen and Kopf (1995: 270) have termed 'ideological dehumanisation', in which the lives of those to be exterminated disappeared behind techno-scientific jargon. Within this genocidal system, labour was compartmentalised so the production, stationing and deployment of nuclear warheads was divided amongst many individuals, few of which could see the bigger picture. Finally, nuclear war entailed, even more than the aerial bombardments of World War II, an act of distanciation, of killing at a distance.

The decision to exterminate would have been taken by people in subterranean retreats, far removed from the sites of extermination (ibid.: 275).

Clearly, the nuclear missile site was a space of extermination. The double barbed wire, watchtowers and lampposts were indicative of a certain conception of hermetically sealed space (see Figure 5.3). As in the Nazi camp system, there was a system of index cards (though not punch cards as in Auschwitz) used to determine the status of its inhabitants – though these inhabitants were not humans but nuclear weapons (US Army 1984: 71). And yet, the tactical nuclear weapons camp was in many ways also rather different from the concentration camp. The former worked as an inversion of the latter in that it turned the security features to the outside: it was not meant to incarcerate but to fortify. In a sense, the nuclear weapons site was an inverted death camp from which the dead bodies had been removed. One of the key problems of the death camp had been the removal of the corpses. As Netz has argued, the logic of the death camp perceived humans 'as future corpses, and so planning was dictated by the problem of disposing of such corpses' (Netz 2004: 221). In the tactical nuclear missile camp, nuclear weapons were occupying the spaces that had previously been inhabited by the concentration camp inmates. It was the survivability of the means of extermination that was at the heart of the camp not the removal of bodies from the body politic. And yet, its purpose was to produce Agamben's 'bare life' – life exposed to death. In this, then, it was in many ways a more efficient space of extermination than the death camp because the levels of potential human resistance had been reduced to almost zero. The guards did not even have to see Agamben's *homines sacri* before unleashing their weapons.

It is important to remember that the guards themselves were of course *homines sacri*, too. Missile silos and tactical missile storage facilities and surrounding villages were of course primary targets for Soviet nuclear missiles, as was recognised in the civil defence literature at the time (no author 1962: 23). This implication of the soldiers as potential victims of nuclear warfare – and this is the third principle, self-destruction – highlights of course the suicidal political impulse inherent in the tactical nuclear missile camp. Transformations in US nuclear strategy during the mid-1950s had 'drastically altered' the rationale for West Germany's frontline soldiers, 'assigning them the task of staving off a Soviet assault long enough for NATO to drop nuclear warheads above them' (Cioc 1988: 9). The West German media had long suspected that the Americans would only release their strategic nuclear missiles once the defence between the rivers Elbe and Rhine had collapsed (Der Spiegel 1967). By the 1980s, West German peace activists and military strategists were in broad agreement on the extent to which the Federal Republic would suffer from a war with the Warsaw Pact. In his 1986 book *Battlefield Germany?*, the military analyst Werner Ebeling worried that the Federal Republic was subject to the highest concentration of soldiers, weapons and military installations within the Western alliance, as well as being the state with the highest concentration of nuclear weapons. He feared that West Germany would be 'the first and "total" aim of conventional and nuclear destruction, the battle-field of a war in Central Europe, with incomparably higher human and material losses than elsewhere' (Ebeling 1986: 31).

This suicidal politics emerged around the West German nuclear weapons silos in a number of ways. Like the strategic nuclear silos located in the United States, Europe's tactical missile sites were of course 'prime targets for attacks by large-yield nuclear weapons or persistent chemical agents' from the Soviet Union (US Army 1984: 19). But, in a more sinister vein, they were also designed to destroy the very land they were designed to protect. This was of course rarely admitted openly. Whilst the West German Ministry of Defence was clear that 'worthwhile targets' were only of a military kind, it did not mention that these targets were likely to be located west of the iron curtain (Bundesministerium der Verteidigung 1956: 10). Given the low-range nature of the tactical weapons stationed in West Germany and the fact that they were meant as 'defensive' capabilities that were to be employed only after an invasion of the Soviet Union, it was clear that these weapons could only be used on German territory. Indeed, in 1964 media reports appeared that NATO was planning to install a belt of atomic demolition muni-tion along the iron curtain (Der Spiegel 1964).

In this way, the tactical nuclear missile camps were spaces of self-destruction that resembled the suicidal policies the Third Reich had tried to enforce on its own population during its last days. As was discussed in Chapter 2, Hitler's policy of scorched earth had been designed to destroy the very territories that had been targeted as the new living space.[6] When it had become clear that the Wehrmacht's retreat was definite, scorched earth had even been extended into the German heartland itself under the so-called Nero command. In continuing with this sui-cidal politics, the nuclear missile sites revealed themselves to be spaces at the vanishing point of biopolitics. They targeted the very territories that were meant to support the state and were meant to destroy anything that could be of use to the enemy. To return to Michel Foucault, these spaces were reminders that 'modern man [sic] is an animal whose politics places his existence as a living being in question' (Foucault 1978: 143).

## Razor Wire and its Discontents

Given its ability to imprint political boundaries onto human flesh, it is perhaps not surprising that barbed wire has attracted considerable interest from political geographers in recent years, particularly from those exploring the enactment of geopolitics at a micro level (Gregory 2004: 76; Megoran 2006) and those exam-ining that paradigmatic space of twentieth-century modernity, the camp. Whilst barbed wire has long been of interest to political geographers and continues to decorate many textbooks in political geography (Dodds 2005; Gallaher et al. 2009; Jones et al. 2014), it has only recently become an issue of conceptual concern.

We have already seen how barbed wire has functioned as a political technology that enables the space of the camp and its politics of expulsion, concentration and extermination. This argument is indeed key to Reviel Netz's (2003) genealogy of barbed wire, which has also informed recent geographical studies of Holocaust films (Carter-White 2013: 21), Israeli settlements in the West Bank (Handel 2014: 506), North American animal migration (Wilson 2015: 7) and theoretical work on camp geographies (Minca 2015). Arguing for an understanding of barbed wire as *the* fundamental political technology of our time, Netz traces modern society's attempts to control mobility and create total space from the nineteenth-century herding of cattle in the American West through the battle-fields of World War I to the concentration of what he calls human cattle in the Nazi death camp.

The case of the tactical nuclear missile camp seems to shine new light on the link between barbed wire as a crucial twentieth-century technology and the prac-tice of extermination. For during the 1970s and 80s in particular, West Germany witnessed a rise of civil unrest and acts of terrorism against US military installa-tions to such a degree that the transportation of nuclear warheads was meant to avoid 'heavily populated areas and those where civil disturbances may occur' (US Army 1984: 56). Arguably, the fact that this insecurity was inflicted by 'a foreign military presence that operated with considerable autonomy' served to increase fears amongst the very populations the Americans had set out to protect (Holmes 2013: 61).

> During this period, many people began to call into question whether the American military presence still served the purpose of protecting Germany, as many local citizens felt that they were in fact being taken hostage by the Americans as they were stationing their most deadly first-strike weapons in densely populated areas of Western Germany. If the Soviets tried to attack the nuclear weapons depots, a substantial part of the civilian population of Germany would be endangered, which would have necessitated retaliation, turning Europe into a nuclear battle-field (ibid.: 36).

Barbed wire, it seems, is a much more ambiguous *symbol* than Netz suggests, and one that has enabled the exercise of sovereign power as much as the resistance to it. It is a very particular material property of barbed wire, namely its *visibility* (its easy photographic reproduction), as well as its *permeability* (its capacity to be seen through), which means it has within it the seeds of its own demise.

Pacifism had always been a problem for successive postwar governments in West Germany. Franz-Josef Strauss, Adenauer's Minister for Atomic Affairs and later to be Minister of Defence, often bemoaned that it was a tragedy that Germans saw only the atom's destructive capacity and not its ability to benefit society. Strauss had been one of the most outspoken political figures behind West Germany's nuclearisation of the late 1950s and early 1960s. In 1955 he claimed that the new nuclear geopolitics needed decisive action if West Germany's 'enslavement' by the Soviet Union was to be avoided. 'To twiddle our thumbs and simply restrict ourselves to passive measures', he claimed, 'would be to commit naked suicide *out of a fear of death*' (Bundestag 1955b: 5610; emphasis added). Strauss was thus implying that only a life that could look death in the eye by playing the potentially suicidal game of atomic war would ultimately be able to avoid national suicide. Indeed, the very idea of deterrence was not too far removed from what Hitler had said in his Second Book, namely that 'Not wanting a war because one has a peaceful disposition certainly does not necessarily mean also being able to avoid it. And wanting to avoid a war at any cost certainly does not necessarily mean saving life from death' (Hitler 1928[2006]: 157). In a way, both Strauss and Hitler were merely reiterating Friedrich Ratzel's aestheticisation of a life that was 'prepared for death'. We are reminded here too of Raskin's insistence that there was 'a necrophiliac quality' to nuclear geopolitics (Raskin 1982: 206). It is thus through the logic of biopolitics that 'modern regimes induce peace simultaneous with war' (Reid 2008: 73).

Clearly, tactical nuclear missile storage camps on West German soil were a crucial site of North Atlantic power and yet they were only a small part of the wider social system that was based on a 'distinct organisation of labour, research and operation, with distinctive hierarchies of command, rules of secrecy, prior access to resources and skills, and high levels of policing and discipline' (Thompson 1982a: 4). Drawing in both military and civilian forces, this nuclear landscape included the nuclear arms industry and its research centres, nuclear test ranges (though not in Germany), early warning systems and a complex network of logistical and communications support for these weapons. Moreover, nuclear weapons were embedded within much larger military landscapes, which included both military bases of all kinds and a much more stealthy redesign of the West German landscape. These landscapes were made up of, for instance, signposts that told a tank crew whether a particular bridge would take their vehicle's weight and motorway segments that could be transformed into *ad hoc* airfields. And yet, any account of Cold War landscapes would have to look in detail at the thanatopolitical space of the tactical nuclear weapons camp.

Like Thompson and Raskin, German anti-nuclear activists were keen to stress the continuities between National Socialist and Cold War geopolitics. Not only had some of the earlier nuclear missile sites been built on old Wehrmacht installations, but the military personnel that took over the command of the new German army in the 1950s had of course fought under Hitler. By the mid-1960s, some more progressively thinking West German generals had become increasingly critical of Bonn's enthusiasm for hardline deterrence and started to see in it a 'strategy to exterminate' that was a direct consequence of Wehrmacht thinking (Bald 2008: 76–7). After a number of German nuclear scientists had complained in 1957 about the Federal Republic's attempts to join the nuclear game, Adenauer invited the scientists to his office and allowed two Bundeswehr generals, Hans

Speidel and Adolf Heusinger, to explain Germany's nuclear policy to them (Cioc 1988: 79). Intriguingly, these men who had been such an active part of the Nazi struggle for *Lebensraum* managed to calm down the nuclear scientists and convince them that West Germany's strategy of hardline deterrence was in fact sensible. Everything was under control.

## Endnotes

1   Soon after having been introduced, Atomic Annie would be replaced with missiles of the 'Honest John' type. The former had been impossible to manoeuvre in daylight as its size posed an all-too-obvious target for enemy planes (Metz 1955: 63).

2   As the chief of staff of NATO's Supreme Headquarters Allied Powers Europe (SHAPE) would explain, 'our mission will be murder and our weapon will be the atomic explosion' – but as a consequence only the Eastern regions of West Germany would be destroyed whilst the rest would survive (Der Spiegel 1955a).

3   For a discussion of the forest's role in the spaces of Nazi biopolitics, see Giaccaria and Minca (2011b).

4   On a more psychological level, too, the Germans were concerned about the resilience of their soldiers. In 1959, the Federal Ministry of Defence warned that the ever more realistic nuclear war games that the US was playing might induce nuclear fear and resignation in the soldiers (Bundesministerium der Verteidigung 1959: 4).

5   Note that German liberals were using similar analogies in the 1950s (Cioc 1988: 44).

6   The soldiers were of course also a threat to themselves. Accidents with nuclear weapons were a recurring phenomenon during the Cold War and the possibility of a soldier or a group of soldiers pushing the button could not be entirely discounted by the two-man rule. As the *Bulletin of Atomic Scientists* reported in 1980, former US servicemen who had worked in nuclear missile camps in Germany had admitted to substance abuse whilst working with nuclear warheads (Dumas 1980: 16).

# Chapter Six
## Enter the Void

*Life itself is at stake in this game*

*Karl Haushofer 1934*

## Nuclear Play

October 1966. Supermarkets and petrol stations are emptied. Migrant workers try to escape West Germany along its motorways as armed insurgencies spread across the young republic. Communist agitators take over the streets, turning peace protestors into rioters, sex workers into spies and foreign grocers into bio-terrorists. In Marienthal, not far from the capital of Bonn, black Mercedes-Benz limousines and army buses pull up in front of a disused railway tunnel. They are chauffeuring the republic's political elites to a safe haven. Behind 25-ton blast doors and more than 100 metres underground lies the West German govern-mental nuclear bunker, a concrete survival pod for the thermonuclear age. The chosen few are greeted by bunk beds, canteen food, stuffy air and neon light. Hunched over strategic maps, they are subjected to an influx of information on political developments and enemy troop movements. As Warsaw Pact forces cross the inner German border, the war begins. It ends only after the North Atlantic Treaty Organisation (NATO) has made use of its first use policy to create a cordon of radioactive rubble on West German territory.

*Cryptic Concrete: A Subterranean Journey Into Cold War Germany*, First Edition. Ian Klinke.
© 2018 John Wiley & Sons Ltd. Published 2018 by John Wiley & Sons Ltd.

Luckily, Fallex 66 (NATO speak for 'fall exercise') was 'just' a game, a hybrid of civil defence exercise and war game that aimed to determine whether NATO members would be able to act smoothly in the early stages of World War III. And yet, the exercise left a bitter aftertaste. Not only had the West German republic's elites called on NATO to deploy nuclear weapons early on in the game, but they had also requested for these to be launched onto West German territory. Somewhat puzzlingly, the West German state had therefore simulated geo- and biopolitical self-annihilation.

This chapter seeks to make sense of this subterranean self-abandonment, ritually repeated in the controlled environment of the nuclear bunker until 1989. In making the case for the fundamentally *unplayful* nature of Fallex 66, I proceed as follows. After unpacking the exercise's script and gameplay, we will zoom in on the East German reception of Fallex 66. As we will see, East Berlin read the war game as a forbidden and obscene enjoyment of self-annihilation in ways that invite a psychoanalytic interpretation of the game. I will therefore flesh out such a psychoanalytic interpretation of geopolitical play and situate this in relation to other concepts of ludic geopolitics. Drawing on this interpretation, I argue that the nuclear war game compulsively repeated the fundamental biopolitical trauma that gave birth to the West German state – the German defeat in World War II and the destruction of German cities by Allied bombing campaigns. It is through this reading that we will begin to understand West Germany's nuclear play as a 'fort-da' game, a term Sigmund Freud (1920[2001]) used to classify his grandson's compulsive habit of abandoning toys only to move them back into view. As in the case of the infant who staged his own disappearance and return in front of a mirror in order to compensate himself for his loss, the West German state playfully participated in the annihilation of its own cities on paper. In this way, the Federal Republic sought to master rather than mourn the experience of urbicide and the subsequent restriction of West German sovereignty by replicating the traumatic experience, but this time assuming the position of the abandoner. It is through this notion of the 'death drive' that the discussion of Fallex 66 is reconnected with the biopolitical impulse for self-annihilation introduced in earlier chapters.

## Fallex 66

Fallex 66 was an 18-day NATO-wide exercise, involving NATO member states, the North Atlantic Council and the major NATO commanders. Played in October 1966, it was the first and publicly most visible of twelve NATO exercises in the Fallex/Wintex-Cimex series, the last of which took place in 1989.[1] These events simulated the transfer of NATO into a state of war, the testing of the alliance's alert systems and the practising of decision-making in real time. In this sense, Fallex 66 presented an opportunity for the West German government to 'influence the actions of NATO forces under the pressure of fast decision-making', as the East German secret service concluded (Ministerium für Staatssicherheit 1966a). The UK Ministry of Defence Chiefs of Staff Committee concluded that Fallex 66 demonstrated 'valuable lessons particularly in nuclear play' (UK Ministry of Defence 1967: A-7).

Rather than involving actual troop exercises, Fallex simulated the interplay of the political elites and the military command in the release of nuclear weapons. Played across the alliance in conjunction with the NATO headquarters, it was therefore a hybrid of a war game and a civil defence exercise. The geopolitical background scenario started with a period of superpower tension during which intelligence procedures and alert measures were practised and was followed by skirmishes ('aggression less than general war') and finally an escalation into general war. Whilst the basic narrative was written by NATO, individual allies would add domestic political events to make the storyline more compelling and challenging for the players. West Germany chose to play Fallex 66 under the most realistic conditions possible – in its governmental nuclear bunker. This subterranean citadel had been meticulously prepared in the runup to Fallex 66 to host 1,200 selected bureaucratic and military staff. This group included not just representatives of the government, the three branches of the Bundeswehr (armed forces), key ministries and federal states, but also the Federal President and the Chancellor (both of whom were played by high-ranking bureaucrats).

Fallex 66 was split up into three continuous parts, 'Top Gear', 'Jolly Roger' and 'Full Moon', only the first of which was publicly acknowledged at the time. During the first part, politicians had not only to respond to a number of domestic political crises but also decide whether to lobby Washington for the use of nuclear weapons. In their nuclear refuge, Defence Minister von Hassel managed to convince the emergency parliament that the use of 'defensive' nuclear weapons, such as the MIM-14 Nike-Hercules, would act to de-escalate the conflict if used on own territory (Dorn 2002: 107). Whilst guidelines on psychological warfare handed out during the exercise made it clear that West German politicians would not publicly admit to the use of such weapons (III. Korps Chef des Stabes 1966), they did tacitly acknowledge their self-destructive logic. In a speech prepared for the nation's darkest hour, the president apologetically explained to his citizens that West Germany's defences included weapons that did 'not entirely lie in the hands of the Bundeswehr' (III. Korps Fernschreibestelle 1966). The first part of the exercise ended rather abruptly with a Soviet retreat in the face of NATO's nuclear threat.

After the politicians had left the bunker, the military stayed on for part two, which simulated full-blown nuclear war and during which NATO detonated atomic bombs both on the territory of the East block and on West German territory that had been taken by the advancing Warsaw Pact armies. The Bundeswehr had already played a crucial role in bringing out the self-destructive tendencies in the first part of the game by requesting the right to forcefully remove refugees from key transport routes (III. Korps Leitungsstab 1966). Only hours after the invasion had begun, it called for nuclear weapons to be released and bemoaned an 'inexplicable timidity' on the side of NATO to set off its nuclear landmines (III. Korps Leitung 1966).When the battle finally arrived, the military furthermore requested napalm to be used against the Warsaw Pact forces, again on West German territory (ibid.). All of this was at a time when West German cities had only been partially evacuated – around 300,000 refugees were still wandering through the most intense fighting zone. The second part of Fallex 66 ended only after NATO had created a 50-km impenetrable cordon of radioactive destruction, from Göttingen in the north to Eslarn on the Czech border (Ministerium für Staatssicherheit 1966a). During the final part of the exercise, the military simulated the post-apocalyptic regrouping and resupply of NATO's remaining forces – 29 days after a first exchange of nuclear weapons. In the course of the entire Fallex exercise a total of 340 nuclear strikes from both sides occurred on the Northern territory of the Federal Republic alone (Northern Army Group 1966: C/2/4).

Fallex 66 was variously described in official sources as an 'exercise', 'play', 'nuclear play' or a 'game' with 'players' (Ministerium für Staatssicherheit 1966b; Northern Army Group 1966, 2-5; UK Ministry of Defence 1967: A-7), and the leader of the West German emergency parliament even called it a 'crap game' (Der Spiegel 1966a: 30). And yet, Fallex 66 in fact only allowed for very limited ludic elements, such as a restricted duration and spatial setting, rules, goals and real-time decision-making. Its gameplay, however, was so linear that it disallowed any ludic complexity or transformative potential. If, for instance, the players decided against a particular (hardline) choice, as happened when they initially refused to let the military remove civilians from the motorways with force, they were subsequently bombarded with messages that outlined an ever worsening situation until they finally gave in. Most importantly, the game did not allow for negotiations with the Soviet Union (codename 'Orange'). It was therefore clear to all participants that the only goal was the smooth release of nuclear weapons rather than the prevention of war.

## A War Game and Its Reception

The German public has known about Fallex 66 since 1966, when details of the exercise were leaked to the press (Der Spiegel 1966a, 1966b; Die Zeit 1966a, 1966b).[2] Interestingly, both these earlier reports and the more recent attention from German bunkerologists[3] (Diester 2009: 179; Diester & Karle 2013: 160, 226; see also Mastny & Byrne 2006: 242) have focused less on the war game's self-destructive logic than on the way in which it enacted the contentious German emergency laws before they had been passed in parliament. Encouraged by NATO itself (Northern Army Group 1966: A/11) and finally adopted in 1968, these laws would transfer powers from the legislative to the executive in an emergency through a simultaneously legal and extra-legal state of exception. Whilst this politically problematic performance generated unrest amongst West German students, it did not prompt a serious debate as to *why* the West German state was playfully abandoning its territory and population in its bunker.

Perhaps unsurprisingly, West Germany's nuclearisation had not gone unnoticed on the other side of the iron curtain. By the mid-1960s, there were a whole range of nuclear weapons systems on West German territory, prompting the East German secret service to speculate that West Germany was now 'more modern and more generously equipped' than NATO's nuclear capabilities in any other part of the world outside the United States (Ministerium für Nationale Verteidigung 1964: 16). Understandably, East Germany had produced extensive intelligence on Fallex 66 from an early stage, which had been presented to the GDR leadership, including Erich Honecker and Walter Ulbricht (Ministerium für Staatssicherheit 1966a).

In such reports and related newspaper articles the exercise was predictably read through the lens of Marxist–Leninist doctrine, and particularly as evidence of Western imperialists' and monopoly capitalists' aggressive plans for world domination (Ministerium für Staatssicherheit 1966b, 1966c: 17). Interestingly, however, there was also a noticeable undertone of enjoyment, even of a sexual kind, in these reports of nuclear self-harm that is uncommon in Cold War discourse.[4] In this way, the Stasi accused the West German political elite and especially the West German Social Democratic Party (SPD) of using Fallex 66 to channel its 'lust for dictatorship' (Ministerium für Staatssicherheit 1966d). Likewise, the official party organ *Neues Deutschland* (1966b) reported that the leadership was participating 'lustfully' in the American nuclear war game and in the liberation of the East. Under the 'passionate' participation of the 'degenerate' social democrats, West German generals were finally able 'to fulfil their dream' of 'pushing the button and starting a nuclear war' (ibid.).

The element of enjoyment became particularly visible in stories of excessive consumption and sexual harassment in the bunker. Whilst these stories may have originated in West German press reports on alcohol consumption, banquets and nocturnal parties during Fallex 66 (Der Spiegel 1966a: 30, 1966b: 27), East Berlin inflated them. A few months later, a spy report on the bunker added a subterranean casino to the story. The casino had, so the report claimed, been designed to entertain the bunker's inhabitants whilst they were deprived from the joys of capitalism (Ministerium für Staatssicherheit 1967: 50). An East German propaganda flyer, possibly produced for circulation in West Germany, claimed that the bunker's corridors and sanitation had been heavily contaminated from excessive drinking and that women had been sexually harassed during the exercise (no author 1966). In effect, the bunker was turned into a space of subterranean luxury, excessive/non-consensual sexuality and forbidden political enjoyment.[5] These stories culminated in a 1967 East German newspaper article that speculated about wild orgies during Fallex 66. 'Against the free flow of sparkling wine and other alcohol', it explained, 'brawls and excessive indecencies occurred' (Tribüne 1967).

These graphic scenes were visualised in a cartoon (Figure 6.1) that is symptomatic of this East German fascination with Fallex 66. It shows a grim-looking West German politician[6] leaning above a blend of a casino board (note the black bow tie) and a strategic map (showing a geographic coordinate system and arrows). He is seen pinning down stylised mushroom clouds whilst a full-blown orgy is underway in the background. Seemingly unbothered by the megadeaths, men are enjoying alcohol, pork (possibly a reference to the greediness of capitalism?) and telephone calls (a symbol of wealth in 1960s East Germany) as well as the company of a scantily dressed woman in a subservient pose. A military officer has joined the party, wearing a Wehrmacht uniform.

Indeed, both in official documents and in newspapers, the exercise was frequently equated with Hitler's final battle and the West German nuclear refuge was compared to Hitler's Wolf's Lair bunker (Ministerium für Staatssicherheit 1966c: 30; Berliner Zeitung 1966; Neues Deutschland 1966a). Clearly, East Berlin was shocked by the way in which West Germany was willing to simulate self-harm and likened this 'terror against the population' to the Third Reich's drive towards self-annihilation (Ministerium für Staatssicherheit 1966d). And yet, it is also clear that East Berlin saw in Fallex 66 an *enjoyment* of Nazi geopolitics

**Figure 6.1**   Cartoon from the East German press (Tribüne 1967). © Copyright by BStU. Reproduced with permission.

and biopolitical annihilation.[7] The East German interpretation thus hinted unmistakably at something traumatic at the heart of Fallex 66, something that had not just to do with the war game itself but with something much larger.[8] Maybe the East Germans, however ideologically blinkered, saw something that the West Germans were perhaps at the time unable – or indeed unwilling – to see.

It is precisely this reading of Fallex 66 as a simultaneously pleasurable and non-pleasurable game of self-annihilation that, I argue, invites a psychoanalytic interpretation that can help us to unlock the fundamental problem of West German geo- and biopolitics. For this form of painful pleasure that is 'alluring yet threatening' is of course at the heart what Lacanian psychoanalysis calls *jouissance* or enjoyment (Kingsbury 2008: 50). In this sense, the concept of enjoyment captures that which is pleasurable (sexually or otherwise) *and* that which is beyond the pleasure principle but which the subject nevertheless strives towards. Enjoyment thus encapsulates something excessive, transgressive and traumatic, something that has the power to convert things into their opposite, the ability 'to render inexplicably attractive what is usually considered a loathsome act' (Žižek 1991: 13).[9]

## Subterranean Play

Whilst the language of games has long permeated both the conduct of and writing about war and diplomacy (Salter 2011a), the politics of play has only relatively recently started to receive attention from geographers (Adey & Anderson 2012; Ciută 2015; Dittmer 2013; MacDonald 2008; Power 2007; Shaw 2010 Shaw & Sharp 2013; Woodyer 2012; Yarwood 2015). In this turn to play, games are increasingly recognised as constitutive of geopolitics itself, the violent mapping and writing of the earth. Seen to extend from the nursery all the way to the war room, scholars have explored ludic geopolitics in geopolitically themed computer games, civil defence simulations, strategy board games, risk management exercises and war toys. This literature has investigated how games fictionalise and make intelligible a complex reality but also how they fuse representational, corporeal and affective dynamics to produce geopolitical order. Perhaps most importantly, play has been investigated in its experimental and potentially disruptive capacities.

The earliest and now sometimes forgotten way in which games entered the study of geopolitics was through a concern with game theory. Whilst war games had always united military strategy and mathematics in an enthralment with the medium of the game (von Hilgers 2012: x), Cold War strategic thinking mastered the art of condensing such war games into mathematical matrices. Schooled in a blend of political realism, geographical determinism and game theory (Dalby 1990b: 175), Cold War strategists read the conflict with the USSR through 'games' like the 'prisoner's dilemma' or the 'game of chicken', in which two opponents are headed for a fatal collision but unwilling to be the first to retreat. Following the logic of this game, the players would deliberately communicate irrational behaviour that did not necessarily reflect actual war plans to scare the opponent into giving up.[10]

Since the early 2000s and the onset of the war on terror, academic debates have increasingly returned to games in an attempt to address the spread of military technology into the most capillary social relations (Lenoir 2000; Der Derian 2001). Such games are increasingly seen to disperse orientalist imaginative geographies amongst gamers (Graham 2006) and rewire the relationship between entertainment, propaganda and army recruitment (Stahl 2010).[11] Simultaneously, geographers started to make arguments about the theatricality and creativity of play. Geopolitics, they implied, is rehearsed, staged and performed, though in ways that often produce unpredicted effects. According to this reading, Cold War militarism was enacted in the most mundane settings though play (MacDonald 2008: 612) and civil defence drew 'directly upon the traditions and techniques of the stage' (Davis 2007: 2). Thus, contemporary simulations, too, are read as materially contingent 'enactments' (Adey & Anderson 2012: 114) that emerge from an affective interplay of performer and audience (Dittmer 2013: 7).

In making such arguments, geographers have increasingly pointed towards the ambiguous and experimental nature of these playful performances. Though cautioning against overly emancipatory readings of games, Hughes (2010: 126) has argued that play is neither pre-determined nor determining. Similarly, Power emphasises that 'games are often used in ways not intended or anticipated by their developers or military sponsors' and that 'data, rules and codes can be modified to produce different narratives' (Power 2007: 282). Adey and Anderson have called on scholars to take seriously the 'creative and surprising or unpredictable and contingent' elements of 'exercise play' (Adey & Anderson 2012: 103; see also Shaw 2013: 793). Some authors have even understood play as a realm of 'fundamental creativity' that holds 'political possibilities' (Shaw & Sharp 2013: 342, 344). This reading draws attention to the disruptive possibilities of play, its ability to obstruct the smooth exercise of power, to question authority and spark political transformation.

The following section suggests a psychoanalytically informed alternative to the emerging debates on ludic geopolitics by emphasising the traumatic, self-destructive, and non-playful dimension of certain games. Whilst nuclear play exhibits some of the characteristics of games, such as rules, players and decision-making, it crucially lacks its disruptive and transformative properties. Fallex 66 neither unlocked the creative capacities of its players nor replaced the normal laws of society in a carnivalesque state of exception. When placed within its political context, it becomes clear that the West German participation in Fallex 66 was more than a mere rehearsal of the Cold War or a performance of deterrence. Instead, it compulsively re-enacted the German experience of urbicide and the loss of sovereignty in the space of the bunker. Although players would have been aware that Fallex 66 was 'just a game', at a deeper level, they were acting out something that was perhaps a little 'too real' to acknowledge.

Whilst psychoanalytic approaches are far from new to human geography (Pile 1996; Social and Cultural Geography 2003), attempts to think political geography and geopolitics through notions of desire, lack, fantasy or libido are mostly rather recent (Müller 2013; Nast 2003). As in other academic fields, psychoanalytic geographies seek to unearth agencies that escape consciousness by interpreting narratives, fantasies and symptoms of 'patients' (be they individuals or social groups) as well as the socio-material settings in which these narratives unfold. Psychoanalytic theory is particularly useful for an analysis of seemingly 'irrational' behaviour and has long developed the conceptual tools to dissect the dynamics of repulsion and desire that surround phenomena like nuclear weapons or the euphemistic and self-deceptive discourses of nuclear deterrence. In this way, I build on Kingsbury's insistence that psychoanalysis offers geographers 'methodological inquiries into the ways in which desire and drive inform socio-spatial relations' (Kingsbury 2010: 530).

In 1995, the late British psychoanalyst Hanna Segal noted in a piece on nuclear war that whenever psychoanalysts tried to apply their insights to areas outside the clinic, there was usually 'an outcry that this [was] not the province of psychoanalysis' (Segal 1995: 191). And yet, she insisted not just that Freud had always been interested in the relationship between individuals and society but also that psychoanalysis was particularly suited to studying the phenomenon of nuclear warfare. After all, nuclear planning and strategy displayed the very self-destructive and irrational behaviour that psychoanalysts had long sought to examine (see also Jacobsen 2013: 17). Segal argued in particular that the bomb aroused 'the most primitive psychotic anxieties about annihilation, and mobilize[d] the most primitive defences' (Segal 1995: 197–8) and that nuclear strategy was 'like a surrealist scenario, an unbearable nightmare or a psychosis' (Segal 1997: 117). Given that the interpretation of collective fantasies or nightmares in films and novels is now well established throughout the humanities, there is arguably no reason why the geo- and biopolitical fantasy of nuclear war should be spared the same treatment.

Indeed, we may add to this that Freud himself was directly concerned with the question of self-destruction, not merely of an individual but of a society, even a species. In *Civilization and its Discontents* he warned that humans had 'gained control over the forces of nature to such an extent that with their help they would have no difficulty in exterminating one another to the last man' (Freud 1930[1989]: 112). A statement often taken to anticipate the holocaust, it also uncannily foreshadows the technological possibility of nuclear extermination. The point in following Freud into his conception of the death drive is not so much to ontologise aggression or to oppose 'aggression' to 'civilisation', as he did in *Civilisation and its Discontents*, but to propose that human subjects repeat non-pleasurable behaviour and that these self-destructive rituals can often be linked back to traumatic experiences of physical and symbolic violence.

Whereas Freud had been convinced that the psyche was governed by the so-called pleasure principle for most of his career, he started to question this earlier conviction in the later stages of his oeuvre as he grappled with forms of destructiveness, which seemed to elide the pursuit of pleasure and the avoidance of pain. Freud was particularly puzzled by non-pleasurable and repetitive behaviour that he referred to as 'the mysterious masochistic trends of the ego' and that he had observed not just in children's play but also in the behaviour of soldiers who had returned from the trenches of World War I (Freud 1920[2001]: 14). He concluded that these neurotic effects were best understood as a return of the repressed, a compulsion to repeat that stemmed from earlier traumata that were playing out in the present.

Freud used these observations to propose his theory of the death drive, positing that next to other (such as sexual) drives, humans were subject to a self-destructive drive for which life was all but a short inconvenience. Freud saw the drives as subject to a different and more complex economy of desire. Unlike instincts (such as hunger or thirst), which are simply directed at satisfaction, Freud's drives encircle objects, thus resulting in 'excessive, repetitive and potentially destructive' patterns (Kingsbury 2010: 520). The paradox of non-pleasurable behaviour is thus resolved by an understanding of how actions that are non-pleasurable to the conscious subject can be pleasurable for its drives, located at the level of the unconscious. In short, the subject keeps putting itself in unpleasant and disturbing situations because it is here that the drive can encircle its object.

Freud famously illustrated his concept of the death drive in his 1920 essay *Beyond the Pleasure Principle*, through the 'fort-da' game, a compulsive game that he had observed in his grandson Ernst. The 18-month-old boy, who had lived in the same house as Freud for a while, had displayed the irritating habit of abandoning toys only to move them back into view. The child repetitively performed this game on a number of objects, including a wooden reel attached to a string, which enabled him to effortlessly ritualise the exercise. He would typically exclaim 'o' when discarding an object (which Freud interpreted as standing for the German word 'fort' [gone]) and a joyful 'da' ([there]) upon embracing it.

Freud unpacked this little game of disappearance and return by contextualising it within the child's relationship with his parents who had praised the boy as particularly well behaved. 'Above all', Freud remarked, 'he never cried when his mother left him for a few hours' (Freud 1920[2001]: 14). Freud held that the departure of the mother, Freud's daughter Sophie, could not have left the child indifferent and that his repetitive game therefore was a return of the repressed (he was not allowed to cry). 'Throwing away an object so it was "gone" might satisfy an impulse of a child's, which was suppressed in its actual life', Freud speculated. But whilst he saw the child's urge 'to revenge himself on his mother for going away from him' (ibid.: 16), another reading would emphasise that the child abandoned itself. Ernst had after all played an even more revealing version of the game in which he made himself disappear in front of a mirror.

Seen in this light, Freud's grandson used the mirror to place himself in the position of his mother. He compensated himself for his great achievement of not crying by staging his own disappearance and return, as seen by his abandoner. In this sense then, the 'fort-da' game was an attempt to 'master rather than mourn' a situation of loss over which he had no control, in this case the disappearance of the mother (Kingsbury & Pile 2014: 4). The game demonstrates both the movement from passivity to activity as well as the constitution of desire through mimesis; the child *wishes* to abandon its passivity for the mother's active position. All of this happens in play. 'To appear in the place of another' is an act of identification that is always already playful 'in the sense in which one plays a role' (Borch-Jacobsen 1982: 35). By playing his mother over and over again, Freud's grandson was effectively caught up in the attempt to constitute himself as an independent and active subject.

## Self-Annihilation

It is in this idea of a subject that simulates – but in effect lacks – autonomy that we can suggest some tentative connections to our West German case. This is not to infantilise the young semi-sovereign republic but rather to suggest that like traumatised subjects, states and their elites, too, can show a compulsion to repeat. It is important to stress that West Germany did not have a veto over the use of tactical nuclear weapons on its own territory – but merely the right to consult the United States in its decision to use such weapons (Buteux 1983: 104). After a Soviet invasion, Bonn would have been at the behest of Washington, much in the same way as Germany's urban population had been at the will of the Allied bombing squads in the final days of World War II.

There are, we can suggest, four further clues that point us in the direction of Fallex 66 as a Freudian 'fort-da' game: the self-destructive and repetitive nature of the game, the element of enjoyment as seen by an external observer and the setting of the game within the simultaneously traumatic and tabooed space of the bunker.

Firstly, Fallex 66 was clearly a suicidal game for West Germany, even though it was not acknowledged as such by the majority of West German commentators at the time. They should have listened to the former Wehrmacht general Günther Blumentritt, whom we have encountered in Chapter 3 and who had in 1952 described West Germany's strategic position as follows:

A sober reflection of our situation shows us that we would have to endure the first flood of Russian forces after an attack. From both sides our fatherland would turn into a playground for the bombers and tanks of both sides. The retreating forces would have to carry out operative destruction. Soon Central Europe would perhaps look like Korea today (Blumentritt 1952: 40).

As we have already seen in previous chapters, other critics spoke of West Germany's approach to nuclear war as 'suicidal defence' (Afheldt 1983: 13) or a play with 'national suicide' (Borinski 1989: 531).

Secondly, Fallex 66 was a repetitive game. As mentioned above, it was only the first and publicly most visible of twelve NATO exercises in the Fallex/Cimex/Wintex series that were ritually played in the bunker until 1989. The vast majority operated with surprisingly similar storylines. After geopolitical tensions in the East block and a buildup of troops along the inner German border, the war would begin. NATO would respond with the selective release of nuclear weapons, often after the use of chemical weapons by the Soviet Union. Most importantly, the majority of exercises simulated the West German request for nuclear weapons. Even during Wintex 85, an exercise that was explicitly designed without the need for atomic weapons, the West German government demanded the use of such weapons, as was widely reported at the time (Der Spiegel 1985b; Süddeutsche Zeitung 1985). This caused a public uproar, to which NATO headquarters responded, apologetically yet somewhat ironically, that the game had 'no relationship to political reality' (Volksblatt Berlin 1985).

Thirdly, East Berlin spotted in Fallex 66 and later games an element of enjoyment. In doing so, it was, much like a psychoanalyst, operating at a partial distance from the game. This meant that it was sufficiently involved to understand the game's deeper historical meaning but external enough to make the crucial observation that the West German state was engaged in something it could only do in the subterranean and forbidden space of the bunker. Whilst East Berlin's reading of the exercise was clearly ideologically tainted and informed by tropes of bourgeois decadence and Western degeneracy, it was also remarkable in its framing of the exercise. By juxtaposing annihilation, over-consumption and sexuality, the East Germans hinted at a distinct combination of pleasure and non-pleasurable enjoyment in rather a similar way to how Freud had observed his grandson's game. The East Germans could see that the kernel of Fallex 66 was precisely the kind of geo- and biopolitics that had turned on itself during Hitler's twelve-year Reich, a form of power that the postwar German state had been denied (Bassin 2003: 361).

Finally, like the original 'fort-da' game, Fallex needs to be understood as a 'profoundly spatial' exercise (Pile 1996: 134); it is about hierarchical relationships that manifest themselves in space (*fort, da*). It is important to remember here that the West German government chose to play Fallex 66 in a very particular space – the nuclear bunker. Whilst there may have been strategic reasons for Bonn to host Fallex in its command bunker, it is important to note that elsewhere the exercise was played above ground. As NATO noted casually in one of its post-exercise reports, play had been conducted from 'peace locations' and 'during extended working hours' (Northern Army Group 1966: 5). We have already noted on a number of occasions the nuclear bunker's function as both womb and a tomb. Indeed, nuclear bunkers are not just geopolitical spaces (i.e. places where geopolitics becomes tangible), but are biopolitical even necropolitical spaces, too.

Nuclear command bunkers are not simply places where sovereign power seeks shelter from the nuclear holocaust outside, but are also the political spaces in which the sovereign decision over survival or extinction is located during nuclear war. By constructing the present as past in an apocalyptic future, nuclear bunkers always already play with an aesthetics of 'disappearance' (Virilio 1975: 57; Vanderbilt 2002: 128) that resembles the compulsive disappearance of Freud's grandson. We can also think of the nuclear bunker as a 'masculine fantasy, participating in an erotics of death that is not subject to self-analysis even as the bombs begin to fall' (Masco 2009: 26) and which as such was rightly described by the West German anti-war movement as grave, coffin or crematorium (Schregel 2011: 193). The bunker was 'an anxious symbol of invincibility and a neurotic defence against the tide of subversive otherness that threatened to invade recently acquired Nazi lebensraum' (Featherstone 2005: 303). As Speer had noted about Hitler's bunkers after 1944: 'If ever a building can be considered a symbol of a situation, the bunker was it' (Speer 1970: 526).

In his excavation of London's highly securitised MI6 headquarters, Pile (2001) argues for an appreciation of the built environment's repressed underside. If we follow this approach, the forgotten site beneath the Federal Republic's nuclear bunker comes into vision, revealing an early history of extermination. We have already seen in Chapter 5 how the West German government had constructed its nuclear command bunker on the very site that had served as an underground slave labour camp for the production of the V2 rocket. Although the V2 is often remembered as a military failure, it did serve as the prototype for both the Soviet and North American ballistic missile programmes, and therefore could be seen as a fundamental technological innovation that heralded the Cold War. The site's reinforced concrete had also, in the final months of the war, given shelter to civilians who were fleeing from Allied bombardments that were increasingly targeting surrounding villages (Bundesinnenministerium 1959). In 1960, this dark history was acknowledged in ministerial correspondence by Theodor Busse (Bundesinnenministerium 1960c). The former Wehrmacht general who had been involved in Hitler's self-destructive battle until the final hour is likely to have written the West German script for Fallex 66, too (Diester 2009: 181). Here we get a sense of the many links and continuities between three types of geographical violence: the Third Reich's genocidal war, the aerial bombing campaigns against Germany, and finally, the destructive logic of the early Cold War.

Much has been written about the role of the Holocaust in the reconstruction of post-1945 German national identity, but whilst few Germans had seen a camp from the inside, many had personal memories of Allied bombing. The destruction of German cities and towns functioned as a publicly visible implosion of the fascist fantasy and its aim of conquering Eastern *Lebensraum*. It was the moment when the genocidal war fell back on the 'chosen' *Volk* itself, and the bunker, rather than the camp, was for many Germans the spatial setting in which this final episode was experienced. In this sense, the bunker was not just a paradigmatic fascist space because of the *Führer's* suicide in his Berlin command bunker, but also because of the large-scale civilian bunker programmes that had mushroomed in German cities during the early 1940s in response to the threat of Allied bombardment. The population's retreat into these wombs/tombs as their cities increasingly turned into ruins was in a sense only the logical consequence of a geopolitical obsession with soil, death and decay.

Nevertheless, and this is the important twist here, the memory of these traumatic events was tabooed in post-war Germany until the 1990s when a debate on the destruction of German cities started to emerge (Zehfuss 2007: 78). Reading the taboo as an attempt to constrain the simultaneity of conflicting impulses, Freud (1913[2001]) argued that particular compulsive or obsessive acts arose in response to self-imposed prohibitions. The German émigré writer W. G. Sebald was puzzled by precisely such a taboo, by the way in which his countrymen had treated the annihilation of their cities as a 'shameful family secret' (Sebald 1999: 10). He was struck by the 'extraordinary faculty for self-anaesthesia shown by a community that seemed to have emerged from a war of annihilation without any signs of psychological impairment' (ibid.: 11). He observed that

> The almost entire absence of profound disturbance to the inner life of the nation suggests that the new Federal German society relegated the experiences of its own prehistory to the back of its mind and developed an almost perfectly functioning mechanism of repression, one which allowed it to recognize the fact of its own rise from total degradation while disengaging entirely from its stock of emotions, if not actually chalking up as another item to its credit its success in overcoming all tribulations without showing any sign of weakness (ibid.: 11–12).

It is precisely this unwillingness to show weakness in the face of traumatic experiences that accompanied West Germany's willing participation in Fallex 66. Whilst West Germany's new Allied guardians actively encouraged the Federal Republic to participate 'as fully as possible' in all stages of nuclear self-destruction (Northern Army Group 1966, C/2), the Germans were eager to show that they were brave enough to do so. At a number of occasions during the game, the Bundeswehr emphasised that neither the 'war of brothers' with East Germany nor the many casualties of nuclear war impressed its soldiers enough to destabilise morale (I. Korps Stab 1966a: 32; 49). Similarly, a fictitious digest of the foreign press that was passed to the players during the exercise praised the German population for its calm response to the threat of nuclear annihilation (I. Korps Stab 1966b).

In the subterranean space of the bunker, the West German elites can therefore be seen to have replayed the state's abandonment of its citizens and the annihilation of 'own' settlements from above. It therefore becomes clear that the bunker, as the very particular *locus* of this fundamentally unplayful game, served to re-enact traumatic experiences in front of the very nations (now NATO allies) that had unleashed the urbicide of cities like Dresden or Cologne. In a sense, West

Germany was enjoying the forbidden fruit of geo- and biopolitics. As members of NATO, the West German elites could at least pretend to have some control over the act of extermination, even though the final decision on the use of nuclear weapons lay with the US president. In the simulated killing of own cities, they had found a stick on a string that gave them an illusion of control. From their place in the bunker, they ironically saw their own towns and villages from the perspective of an Allied bombing squad – as a toy that they could make disappear on a map.

Writing in 1985, the psychoanalyst Carl Nedelmann noted the way in which Anglo-American defence experts knew Germany only as the 'nuclear theatre' during most of the Cold War (Nedelmann 1985: 17). Nedelmann suggested that the Federal Republic's elites knew this very well and were suffering from a sense of inferiority vis-à-vis its Allies that was further increased by the failed attempts to acquire its own nuclear weapons during the chancellery of Konrad Adenauer (ibid.: 21). Drawing on the psychoanalytic insight that the failure to achieve goals tends to give rise to emotions of inferiority and guilt, Nedelmann claimed that West Germany's sovereignty deficit functioned as the 'open wound' of Germany's national identity. The feelings of inferiority and guilt that became so characteristic of postwar German nationalism may have therefore been rooted not merely in the genocidal war that Germany had inflicted upon the world but also in its failure to have achieved great power status after the war.

Ultimately, Fallex 66 was a very particular type of game because it incorporated within it the point from which the game can no longer be replayed. As Hanna Segal noted, nuclear warfare had made it possible to 'annihilate the world, but not reconstruct it' (Segal 1995: 197). In other words, the two blocs' ever expanding nuclear arsenals made the 'fort' increasingly easy to achieve, whilst the reality of nuclear overkill was making the possibility of the 'da' vanishingly small. This 'world of primitive omnipotence', Segal held, was governed not by the fear of death but 'by wishes for annihilation of the self and the world, and the terrors associated with it' (ibid.). And yet, nuclear play, to use that euphemism one last time, was maybe not all that different from Ernst's 'fort-da' game. For although the arms race was making a 'replay' highly unlikely, emergency planners were of course fantasising about the world's repopulation after a nuclear war. Indeed, NATO continued to quite casually refer to the final part of Fallex 66 as a 'post nuclear setting' (Northern Army Group 1966: C/2). It is this insistence on the possibility of a post-apocalyptic world that shows the limits of the death drive even within the fantasy of nuclear Armageddon.

## The Death Drive of German Geopolitics

Like Friedrich Ratzel, Karl Haushofer was concerned with what remained of a nation's culture after it had turned into 'ruin landscapes' (1934a: 96). As we have seen, he would contemplate the possibility of the German nation's 'complete annihilation' even before the advent of the atom bomb (Haushofer 1944: 635). At a time when Germany had increasingly become the target of Allied bombing campaigns, he began to fear that the destruction of German cities entailed the end of Germany as a *Kulturvolk* ('civilised nation') and that its cultural achievements would be forever lost. He was clearly afraid that the annihilation of Germany through aerial warfare would be so total that there would be none of Ratzel's ruins left for posterity, no artefacts to remember the Germans by. So when Haushofer wrote in his 1946 suicide note that he neither wanted a state nor church funeral nor indeed an obituary, this should be read as an attempt to rid himself of that which he thought to be the highest form of life – the preservation of traces of life beyond death. By being, as he hoped, 'forgotten and forgotten', his life would be disallowed ruination (Haushofer 1946).

In a note entitled 'On death' and written in 1903, only a year before his own death, Friedrich Ratzel mused on the omnipresence and ontological 'easiness' of death (Ratzel 1941: 302). 'No mountain range or wall', he felt, stood between life and death, and pondered that 'the way into the great and dark gateway' was really 'rather flat'. Ratzel romanticised the experience of witnessing another human die and described that brief moment of transition from 'time to eternity' admiringly as a 'sanctification' (ibid.: 303). Rather than tabooing death, Ratzel felt, humans needed to embrace it. 'The more we become accustomed to death', he wrote, 'the smaller the barriers to eternity will become'. He concluded that 'the more life is prepared for death, the more beautiful it becomes' (ibid.: 302). Virilio has noted that it was the destruction unleashed by the Allied air raids that had first made Albert Speer realise that 'when everything was destroyed, in the greatest dearth, the institutions continued to function' and thus 'a social system survived' (Virilio 1975: 58).

It is this drive to preserve something by pointing it towards death, crystallised in Speer's theory of value in ruins, that links up the psychoanalytic concept of the death drive, on the one hand, with the thanatophilia inherent in fascist (geo) politics, on the other. In this interpretation, the Cold War was no more than a compulsive repetition of this most fundamentally biopolitical move – the insertion of death into life.[12] The security of the state and its functions could only come at the expense of the security of the population. It was precisely because the West German government had the possibility of retreating into the sanctuary of its command bunker that a nuclear war in which the West German population would be wiped out became 'rational' at all.

It is important to understand that Bonn's willingness to destroy and radioactively contaminate its own territory in war games like Fallex 66 showed not just a determination to discourage a Soviet attack but also a compulsion to self-harm. In a sense, West Germany was only acting out what Hennig had in 1935 described as the necessity to 'incapacitate' states that had become 'unfit for life [*lebensuntüchtig*]' (Hennig 1935: 90). Rather than a mere enactment of Cold War deterrence, Bonn's war game was a re-enactment that compulsively repeated the fundamental bio- and geopolitical trauma that gave birth to the young West German state: the annihilation of German cities through Allied bombing campaigns and the restriction of West German sovereignty. In this way, the Bonn Republic abandoned its dwellings in a struggle against 'totalitarian' Communism in much the same way that the Allies had sacrificed those same towns and cities in the struggle against fascist totalitarianism two decades earlier.

Whilst Fallex 66 did indeed display ludic elements, the war game was devoid of the very ingredients that make a game playful, such as spontaneity, disruption and creativity. At the heart of Fallex was not playful experimentation but a self-destructive compulsion to repeat a trauma caused by an experience of loss. If it was a game, then it was only a 'fort-da' game in which the semi-sovereign West German state transposed the unequal power relations between its new parental allies and itself onto its relationship with its own civilian population – only this time simulating control of the situation. The game's true aim for West Germany, then, was to master rather than mourn the experience of urbicide and bunker life.

Similar attempts to master the trauma of annihilation could also be found in literary texts. The 1957 novel *Es geschah im Jahre 1965* ('It happened in the year 1965'), for instance, reveals like few other documents a possibility of thinking the Cold War through the political lens of the previous war. In that, it reveals the fundamental compatibility of National Socialist ideology with the biopolitics of nuclear war. Written by the former Nazi writer Edwin Erich Dwinger, the novel follows a number of German, American and Soviet protagonists, most of whom are soldiers, through World War III. The war begins with a Soviet nuclear attack on NATO and ends with the complete annihilation of the Soviet Union with nuclear and biological weapons and, somewhat miraculously, German reunification and the foundation of a United States of Europe. Interestingly, Dwinger's novel is not just saturated with racist language and a very familiar idealisation of the German peasant, but it is also permeated with the mythology of geopolitics. Throughout the text, events in the lives of his protagonists are placed in their larger military strategic and world political context. At one point, the positive effects of geographical space on a nation's resilience in nuclear war are explained; elsewhere the reader learns that only subterranean civil defence can convince the Soviet enemy that Germany 'cannot be exterminated biologically' (Dwinger 1957: 66; see also Lemke 2007: 79).

Many of the novel's scenes take place in bunkers. By the mid-1960s, West German cities have been turned into large subterranean landscapes, with shopping malls, sweet shops and restaurants. For Dwinger, who favours an agricultural utopia, these underground cities are dystopian spaces. In one scene, one of his protagonists finds himself glancing into the eyes of a mannequin, which he has spotted in the window of a subterranean clothes shop, only to ponder about its numb and empty facial expression. We get the feeling here that American consumer capitalism, symbolised too by the squeaky, syncopated strangeness of jazz music, is but a distraction from the existential struggle that characterises the nuclear age. Although Dwinger's story is favourable both towards NATO as a military alliance and the notion of the 'Occident' as a civilisational category (which he has set up against 'Bolshevism'),[13] the storyline does allow for a more nuanced interpretation. Crucially, whilst Germany emerges from the nuclear apocalypse strangely unscathed, the United States has lost many of its cities and the United Kingdom has even been completely annihilated alongside the Soviet enemy. Dwinger has difficulties concealing his enjoyment at describing the destruction of New York – and the liberal world order that the city stands for – by Soviet nuclear missiles:

> The first thing that fell was the United Nation's palace, this fairy-tale building of twinkling glass. It collapsed into the Hudson river like a house of cards, almost symbolic for its failures, causing an immense tidal wave. Wall Street is now no more than a pile of rubble [...] The first National City Bank, the Irving Trust Company, in between them the tiny Trinity Church, [...], the Empire State Building and the Rockefeller Centre... none of these behemoths in stone has survived (Dwinger 1957: 134).

Thus, by saving only Germany – and not its Allies – from nuclear Armageddon, Dwinger's narrative functions as an inversion of the narrative of Fallex 66 and similar nuclear war games, in which Germany was the *only* battlefield of World War III. In this, then, we can speculate that these narratives may hide a deeper fantasy. At a subterranean level, Dwinger seems to desire not just the urbicide of Allied cities but to survive total war – precisely what the Third Reich failed to accomplish. This is, in other words, a fantasy not just of destroying the Soviet enemy in the East but also the 'parental' Allies in the West. Interestingly, Freud remarks in *Beyond the Pleasure Principle* that when little Ernst was five and three-quarters, his mother died. Perhaps unsurprisingly, Ernst 'showed no signs of grief' (Freud 1920[2001]: 16).

## Endnotes

1  Not all of these have been declassified in the Federal Archives, but extensive intelligence on the later exercises is available at the Federal Commissioner for the Stasi Archives in Berlin.
2  Fallex 62, a 1962 exercise, had famously led to a scandal after the magazine *Der Spiegel* had revealed the country's lack of preparation for war (Der Spiegel 1962). The state subsequently attempted to shut down the magazine, but backed down in the face of mass demonstrations.
3  For a discussion of the practice of bunkerology see Bennett (2011a).
4  Note that whilst Ó Tuathail and Agnew (1992: 200) have pointed out the sexual undertones of early Cold War American geopolitical discourse, their work did not include the dimension of self-harm.

5  During a similar exercise in 1975, the East German press construed a knife attack on a female secretary not as attempted murder but as a sexual assault (Berliner Zeitung 1975).

6  This was probably the Christian Democrat Minister of the Interior at the time, Paul Lücke.

7  Gregory (2011: 261) notes that whilst the term annihilation is today usually reserved for the holocaust, it was also frequently used in British and North American discussions of the Allied bombing campaigns during the 1940s and therefore has a wider function within discourses of total war.

8  The controversies that emerged around Fallex 66 and similar exercises in West Germany would suggest that West Germans too were uneasy with these simulations, but they did not go so far as to insinuate a subterranean enjoyment in self-harm.

9  It is interesting to point out that in 1966 the GDR had started planning a new governmental nuclear bunker of its own that roughly resembled its West German counterpart in design and function (Diester & Karle 2013). It also included similar, though less elaborate, simulations that played through nuclear war (Freitag & Hensel 2010: 153). Despite the East German regime's seeming repulsion by the events in West Germany's governmental bunker, it was nonetheless engaging in very similar games. We are instantly reminded of Freud's notion of emotional ambivalence; the idea that within love is hostility in the unconscious and vice versa (Freud 1913[2001]: 70; see also Beck 2011: 83).

10  If known to the enemy, war games like Fallex 66 could of course play a part in the practice of deterrence (Geenen 2006: 51).

11  From this perspective, war games might appear as the true models for commercially produced games like *Diplomacy*, *Risk* or *Civilization* (Salter 2011b).

12  For a perhaps more radical take on this question, see Mbembe (2003).

13  At the end of the war, a NATO general explains that the 'Occident would have been lost, every last square meter annihilated, had it not found together through NATO' (Dwinger, 1957: 294). A German officer explains that Germany would have been lost without tactical nuclear missiles (ibid.: 308).

# Chapter Seven
## Conclusion

## The Nuclear Present

To witness the active remains of West Germany's nuclear biopolitics today necessitates a journey to Büchel, a small village of around 1,000 inhabitants in the rural Eifel region. It is a rather gloomy October afternoon and thick clouds hang over the houses and farmyards. Maybe it is in this bleak atmosphere that the Stars and Stripes that welcome visitors upon their arrival are so noticeable: there are two US flags in front of an old petrol station, a third on a local music shop and a fourth at the entrance to Büchel air base, home to the German Airforce's *Taktisches Luftwaffengeschwader 33* (Tactical Air Force Wing 33). It is here, 8 metres below the earth's surface, that the North Atlantic Treaty Organisation keeps its remaining thermonuclear weapons in Germany, the air-launched B61 bombs, currently being modernised.

The site is surrounded by barbed wire fence and a succession of bunkered fortifications. The air is thick with a biting smell of kerosene. The security guards at the main entrance casually inform guests not to take photos from close up and the local baker smiles uneasily when asked about the military base. Signs warn visitors that soldiers may use their weapons if one tries to enter the site. A herd of deer are unimpressed by a jet plane soaring into the sky, but upon my arrival, they scramble into the bushes.

*Cryptic Concrete: A Subterranean Journey Into Cold War Germany*, First Edition. Ian Klinke.

There seems to be something of a collective amnesia around nuclear war in contemporary Germany. Indeed, many nuclear storage sites have been abandoned since the end of the Cold War and turned into 'no-man's lands' (Leshem & Pinkerton 2016: 44). Others have been transformed into art galleries, regional airports or solar energy complexes. Others again have been reused to store explosives or house asylum seekers. Camp Pydna near Kastellaun, named after the 2nd-century BCE battle of Pydna, now hosts an annual techno festival. Like the ravers, the military seek to reimagine the nuclear weapons storage site as a happy place. Büchel air base's friendly website informs the public that soldiers regularly meet local residents about the inevitable noise pollution and help to organise a whole range of festivities, which seek to connect the local community and 'their air base'. We learn that the site's commander is a man called Andreas Korb ('married with two children') and that his Tactical Air Force Wing 33 has ensured 'peace in Europe' since 1958 (Luftwaffe 2011, 2014).

Elsewhere, the language has remained a little more geopolitical. Jamie Shea, NATO's Deputy Assistant Secretary General, is famous in Brussels for his rhetorical skill, expressive body language and London twang. In 2014, he came to University College London to talk to students about the Ukrainian crisis and Europe's new security architecture. Charming his audience with jokes and anecdotes, Shea told students the story of an underfunded alliance that urgently needed to up its military game if it wanted to stand up to Vladimir Putin's Russia. He was excited about NATO's new 'very high readiness joint task force' and about new and larger planned NATO manoeuvres in Eastern Europe. One topic he had preciously little to say about was the alliance's nuclear policy. When prompted to speak about the current modernisation of the B61 bombs in Germany and elsewhere, Shea explained that whilst the life extension programme had originally been designed to provide NATO with a bargaining chip for future nuclear disarmament talks with the Kremlin, Russia's military involvement in Eastern Ukraine had fundamentally altered the strategic context. The B61 was now 'once again a part of NATO's deterrence posture'. 'I'm not nostalgic for the Cold War', he laughed, 'but you have to prepare for Cold War mark two, even if you don't want it'.

As Flockhardt (2013) reminds us, NATO remains something of a 'nuclear addict' – it hangs on to its nuclear weapons despite frequently declaring its desire to abandon them. And whilst the nuclear weapons in Büchel continue to be unpopular in parts of the German press (Der Spiegel 2013; Die Zeit 2016), there are an increasing number of voices in Germany that evaluate the role of nuclear weapons more positively. In January 2017, after the potential NATO sceptic Donald Trump had become US president, the German state TV programme *Panorama* hinted at the possibility that Berlin might now – finally – be in a position to acquire its own nuclear weapons (Panorama 2017a). Panorama was not alone in floating the nuclear option. The idea was also being considered by the daily newspapers *Tagesspiegel* (2017) and *Wirtschaftswoche* (2017). Around 60 years after Adenauer had famously denounced the Federal Republic's ambition to gain what Hitler had failed to achieve, the German bomb was again being debated. Interestingly, however, parts of the German population reacted so angrily to *Panorama*'s proposition that the programme makers had to publish an apology for this advance (Panorama 2017b).

For large parts of the social sciences, the question of nuclear warfare was long considered either boring or passé. As Bruno Latour told his readers in his 2004 piece on the end of 'critique':

> Threats might have changed so much that we might still be directing all our arsenal east or west while the enemy has now moved to a very different place. After all, masses of atomic missiles are transformed into a huge pile of junk once the question becomes how to defend against militants armed with box cutters or dirty bombs (Latour 2004: 230).

Notwithstanding Latour's assessment,[1] it is perhaps interesting to note that missiles and bunkers have not disappeared from the Pentagon's radar. It is not just that recent years have seen a return of a Cold War framework amongst US journalists, think-tankers and policy-makers. Rather, the Pentagon remains concerned with the invulnerability of enemy bunkers, prompting the US government to invest 'heavily in a new generation of sensors designed to radically reduce the opacity of the ground' (Graham 2016: 344). Indeed, one programme, baptised 'Transparent earth', 'aims to offer a Google Earth-like 3D interface that would be able to display the physical, chemical and geological properties of Earth to a depth of five kilometres' (ibid.). This 'dream of a nuclear bunker buster to end all bunker busters' remains a crucial fantasy for contemporary military strategists, as Bell (2008: 224) has argued.

In an age of widespread anxieties, subterranea continues to be alluring. In 2015, Vivos, a Californian bunker construction company, announced that it intended to transform a former East German bunker near Rothenstein into a survival capsule for the super-rich. Wealthy families, they promised, would be able to endure a nuclear war or natural disaster in blast-proof chambers connected by over 3.1 miles of tunnel. *Forbes* magazine was quick to describe the project as 'an invitation only, five star, underground survival complex, similar to an underground cruise ship for the elite' (Forbes 2015). But there was more at work here than the logic of late capitalism, for Vivos, with obvious biopolitical undertones, also promised to house a 'collection of zoological species' and a 'DNA vault to preserve and protect the genomes of millions of donors' (ibid.). Interestingly, the German press noted that this particular bunker would have difficulties getting planning permission (Thüringer Allgemeine 2015; Die Welt 2015). And yet, what mattered to Vivos was perhaps not the actual construction of the bunker but merely that the story was circulating in the transatlantic media landscape, thus attracting interest from potential clients. This illustrates perhaps that

> bunkers are places most of the time encountered more in fiction than reality, simultaneously figured in popular culture as a place of ultimate control and a place of abject defeat. Thus, in many ways 'the bunker' is a phenomenon for which we know more about the image than the reality (Bennett 2017: 7).

In attending to the fantastic character of the bunker, however, we should not be discouraged from engaging with the actual ruins of Cold War bunkers and the geo- and biopolitical context in which they were embedded. As we have discussed in previous chapters, nuclear missile sites, too, were architectural spaces that articulated both a geopolitical context (the Cold War) and a biopolitical concern with survival and extinction. Of course, the tactical nuclear weapons stationed in West Germany were only part of the West's nuclear arsenal. In the United States, intercontinental missile launch sites, such as the Minuteman silos, would reshape the landscape profoundly in the 1960s. And yet, in their subterranean nature (the silos were sunk 80 feet underground), such weapons facilities similarly articulated a fetish with the earth. As Heefner notes,

> At any given moment during the Cold War, no fewer than 200 Air Force officers lived underneath the Great Plains. Their job was to sit, and wait, and hope that the job they had been trained to do would never have to be done. The officers rotated through four-day shifts of 12 hours underground, 12 hours above. They ate below ground. They slept there. A toilet was furnished for their comfort. The necessities were provided. It was something of a submarine under the soil, only one that would never surface (Heefner 2012: 111).

As we have learned in this book, we should not just read subterranea as a biopolitical space of protection and salvation but also as one from which extermination was planned and rationalised.

This reading of the violent landscapes of the Cold War as geo- and biopolitical has crucial implications for how geographers think about the relationship between war and peace. Noting the role of cartography in modern warfare, Edward Said once characterised geography as 'the art of war', an art that could be transformed into an act of resistance if supplied with 'a counter-map and a counter-strategy' (Said 1995: 416). Since Said's musings war has indeed become a crucial area of geographical concern, as witnessed by a volley of books that have appeared over the last decade and a half (Adey, Whitehead & Williams 2013; Cowen & Gilbert 2007; Elden 2009; Farish 2010; Flint 2005; Fregonese 2012b; Gregory 2004; Gregory & Pred 2007; Graham 2004b, 2016; Ingram & Dodds 2008; Macdonald, Hughes & Dodds 2010; Smith 2003; Toal 2017). And whilst this body of scholarship has arguably produced one of the most critically minded literatures on modern warfare available in the social sciences today, it has increasingly been suggested that its focus on warfare and killing obstructs a necessary examination of the geographies of peace (Bregazzi & Jackson 2016; Fregonese 2012b; Herb 2005; Megoran 2011; Williams & McConnell 2011).

Diagnosing critical political geography with a tendency for an agonistic understanding of politics, Bregazzi and Jackson (2016: 3) ask, for instance:

> Why do we overwhelmingly focus in our critical approaches to politics on confrontation, while the innervating force of the demand for justice, and so the constitutive grounds of peace – love, care, compassion, empathy, etc. – are framed as either apolitical or irreducibly infused with their opposite?.

'[L]ooking for violence and calling it critique will not reveal peace', they continue. These sentences have a familiar ring to them. It was during the Cold War that the academic field of peace studies gained prominence precisely as an alternative to the prevalent realism characteristic of war studies and international relations. The rejection of critique as the uncovering of violence echoes moreover a recent disillusionment amongst human geographers with the very notion of critique and a preference for more affirmative forms of political engagement (see McCormack 2012; Woodyer & Geoghegan 2012 and, again, Latour 2004). Indeed, much is to be said about the need to take more seriously the politics of care, compassion and empathy. And yes, a preoccupation with nuclear weapons and military strategy can bring critical geopolitics dangerously close to the analytical territory occupied by classical geopolitics.

Nevertheless, many of the recent calls for a geography of peace are based on an understanding of war and peace as essentially separable phenomena: it is possible to study either war *or* peace. Such a binary way of thinking, however, throws up not merely the analytical problem of distinguishing between war and peace. Rather, such a binary can obstruct a view of the violence *inherent in* discourses and practices of peace, as Amy Ross (2011) has reminded us (see also Koopman 2011). The Cold War – or what strategists have caringly called 'the long peace' – works as a case in point. For we are faced here with a period in history that is understood by many, especially in Europe, to have been *the* success story of pacification. Indeed, in 2012 the European Union was awarded the Nobel Peace Prize for this endeavour (Mamadouh 2014), yet the idea that the European Union is a successful alternative to the politics of war only works if one detaches its emergence from the geopolitical and highly belligerent circumstances of the Cold War (Klinke & Perombelon 2015). As I hope to have shown in this book, the very societies that were 'pacified' after 1945 never managed to remove the ideological and physical architecture of a will to exterminate. Nuclear war relied on the very same infrastructure and materials that were leading to Western Europe's economic recovery: nuclear power stations, motorways and so on. By reaching into classrooms and homes, the Cold War tried to touch the bodies and minds of civilians and soldiers, children and strategists alike. The violent patriarchy of the nuclear family was reproduced at a different level by the military authoritarianism of the nuclear state. And yet, many citizens were oblivious to the degree to which their seemingly peaceful societies were in fact economically and politically geared towards nuclear war. In this, then, the Cold War's violence was both 'intimate' and curiously distancing (Pain 2016).

We may be left wondering whether it is not possible to find within the violent geographies of nuclear war the very things that an affirmative politics of peace is meant to explore. Not only did politicians on both sides of the iron curtain tend to speak of peace when they in fact meant war, but the very material architecture of nuclear war relies on the attributes that contemporary geographers associate primarily with peace: friendship, care, compassion and love. Whilst military alliances like NATO were often written and spoken about in a language of *friendship*, civilian bunkers were meant to materialise the state's (arguably fake) *compassion* for its population. Whereas nuclear weapons were the subject of custodial *care*, nuclear strategists became *emotionally* and sexually *attached* to their weapons (Cohn 1987). In this sense, then, the geopolitical myth of the 'long peace' is an important reminder of what Derek Gregory (2010: 155) has referred to, in a different context, as the 'deadly embrace of war and peace'. Surrounding the geopolitical architecture discussed in this book was an intimate economy of care that protected the very destructiveness of nuclear war. And whilst this economy of destructive care survives into the present, it also lies in ruins.

## Ruin Value

The study of nuclear bunkers and missile sites joins a long list of other geopolitical spaces that have been addressed by political geographers over recent years, from borders (Amoore 2006; Sparke 2006) to airports (Adey 2008), from refugee camps (Ramadan 2013) to hotels (Fregonese & Ramadan 2015) and from sports-utility vehicles (Campbell 2005) to pipelines (Barry 2013b). Some of the spaces seem static, others more mobile or fluid. Others again might in fact be amalgamations of different architectural archetypes. In this way, we might think of the nuclear submarine, that stealthiest of killing machines, as both a submarine living space and a fusion between the nuclear bunker and the missile silo. What emerges from this, therefore, is the necessity of thinking of *Lebensraum* not as a particular kind of space but as a particular *political logic* that can be injected into a whole range of architectural and discursive spaces. Indeed, from the barbed wire fantasy of a Fortress Europe to that of colonising outer space, *Lebensraum* seems alive and well today (Virilio 2012: 45, 67).

This book has sought to show how West Germany's preparation for nuclear war happened, both ideologically and architecturally, in very similar registers to those used by the Third Reich in its socio-material construction of *Lebensraum*.

We have examined the return of geopolitical discourses in the 1950s and seen how the idea of national survival was to be accomplished by West Germany's participation in NATO's nuclear geopolitics. We have taken a closer look at the Federal Republic's obsession with both the soil and hermetically sealed space, which, when fused, resulted in the attempt to secure its sovereignty, limited as it may have been, in subterranea. Finally, we have sought to make sense of the self-destructive (thanatophile) nature of nuclear war games, which the West Germans actively engaged in. We have thus concluded that a fixation on spaces of survival and extermination was not limited to the early 1940s but managed to re-emerge in the planning for nuclear war during the 1950s and 1960s. As in the conceptual universe of Friedrich Ratzel, Rudolf Kjellén and Karl Haushofer, the state was again read in West Germany's early Cold War as an organism, a life form that struggled for survival in a competitive environment and was therefore subject to the forces of natural selection.

At the heart of the analysis were two architectural archetypes – the missile silo and the nuclear bunker. Whilst the architecture of these two spaces is not exclusively German, for bunkers and missile sites existed elsewhere in abundance, the book hopes to have shown an alternative 'German' interpretation of the emergence of the Cold War. Rather than being framed and legitimated through the metaphorics of the American frontier or indeed consumer capitalism, the nuclearisation of Germany happened as a reiteration of project *Lebensraum*, first articulated by Friedrich Ratzel at the turn of the previous century. In this process, Ratzel's geo- and biopolitical logic was often inverted and distorted, but it was nevertheless alive. In this sense, the autarkic space of the nuclear bunker should be read both as a symbol of the Cold War and as the final retreat for German geopolitics.

Nevertheless, it is important to clarify that the book's aim was not simply to compare – or worse *equate* – the Third Reich's genocidal politics with the (luckily) merely hypothetically genocidal project of the Cold War. To do so might not have just run the risk of trivialising the former, but might also have glossed over the important discontinuities that I have traced here. Whereas the Ratzelian conception of *Lebensraum* had stressed the importance of open territorial spaces, the Cold War complemented this more consistently with a focus on small subterranean living spaces, injecting the earth with a new meaning. Whilst tactical nuclear missile camps were most certainly spaces in which extermination became conceivable, the production of death rested on a different relationship between citizen, soldier and corpse. And finally, whilst the ruin aesthetic did live on in the fantasy of repopulating a radioactively contaminated ruin landscape, West Germany's postwar preoccupation with ruins was shaped by the very real experience of urbicide rather than simply by Romanticism.

I hope to have shown that both National Socialism and its attempt to forge *Lebensraum*, on the one hand, and the West German state's attempts to create *Überlebensraum*, on the other, had common conceptual and political origins.

This is to make a case for the fundamental *compatibility* of the logic of nuclear war with the ideas that were first developed by German geographers at the turn of the previous century and that so notoriously left their mark on National Socialism. Much as in the early 1940s, this geo- and biopolitical impulse was directed in the 1950s and 1960s towards the home nation, as the Federal Republic increasingly toyed with a politics of self-extermination. Singling out the German case is thus not to suggest that this play with national suicide was in any way ethically more problematic than that of annihilating enemy cities. Instead, it is to appreciate nuclear self-annihilation in its own right, as a biopolitical vanishing point.

The key to an understanding of how the self-destructive logic of geo- and biopolitics could have re-emerged in Germany (of all states) lies perhaps in the stealthy and subterranean resilience of *Lebensraum* amongst strategists and civil defence planners, both as a discourse about the state and as a way of expressing this discourse through the built environment. Cold War geopolitics is often seen to be devoid of biopolitics, constituting a kind of geopolitics that retained a fascination with military power, maps and hard-nosed decision-making but that had otherwise rid itself of a Ratzelian understanding of the state as organism. And yet, as we have seen, the Cold War state *was* of course engaged in a game of survival and extermination, precisely the sort of competitive game that would have made great sense to someone like Ratzel, given his theoretical preoccupations. Biopolitics is thus, as Derek Gregory reminds us, remarkably resilient. Whilst the biopolitical lexicon has certainly transformed since Kjellén, it has remained a key referent because of its ability to 'make military violence appear to be intrinsically therapeutic' (Gregory 2010: 277).

What, then, are we to do with the biopolitical ruins of Cold War West Germany? In a way, of course, the idea of learning from history is banal. Kissinger argued in his 1957 discussion of nuclear war that 'one of the most difficult challenges' for any nation-state is 'to interpret correctly the lessons of its past' (Kissinger 1957[1968]: 16). Even Hitler wrote in his *Second Book* that it was 'often disturbing to see how little people [were] willing to learn from history' (Hitler 1928[2006]: 38). And yet, we may legitimately wonder whether there is any value in the hypothetical ruins of the Cold War's nuclear geopolitics?

Masco (2008) has suggested that no nation has obsessed as much about ruins as the United States of America. In the United States, images of ruins have played an important political and cultural role.

In the early Cold War period, ruins bec[a]me the markers of a new social intimacy grounded in highly detailed renderings of theatrically rehearsed mass violence. The intent of these public spectacles – nuclear detonations, city evacuations, duck-and-cover drills – was not defence in the classical sense of avoiding violence or destruction but rather a psychological reprogramming of the American public for life in a nuclear age (Masco 2014: 47).

Nevertheless, as I have argued above, it was Germans like Friedrich Ratzel and Albert Speer who first placed the ruin as a central chronotope within a geopolitical imaginary. Indeed, even after World War II, West German houses were imagined by civil defence planners with their ruined state in mind. It is thus no coincidence that *Ziviler Luftschutz* would describe with overwhelming admiration a 'city of ruins' that had been constructed by the Danish civil defence service for training exercises, thus enabling a more embodied experience of nuclear war (Schützsack 1960). And yet, unlike in the United States, ruins were not simply fantastic in West Germany. Whilst in the US, the atmosphere of war on the home front had to be created with the help of cinematic nuclear ruins, West German civil defence could build on the actual experience of aerial bombing during World War II.[2] The traumatic familiarity with aerial bombardment meant that – despite the taboo around the issue – West German civil defence had at its disposal a powerful set of affects, however difficult they were to harness.

For it is important to note that at no time was this manipulative mobilisation of past traumas uncontested in West Germany (see also Biess 2009: 222). On the one hand, civil defence brochures would urge their readers not merely to remember the traumatic episodes of World War II but also the positives, such as the many helpers who had risked their lives during the Allied air-raids, helping the injured and buried persons (Bundesinnenministerium 1958b). On the other hand, however, it was precisely the memories of aerial bombardment that repeatedly mobilised parts of the population *against* the nuclear state. I have not spoken much about these resistances here. This is not because I believe them to be historically insignificant or because I wish to ignore the politically progressive struggle that emerged through these protests, a history that has been well documented by historians such as Schregel (2011) and Nehring (2013). Rather, I hope to have shown that by concentrating on the naked thanatopolitical workings of the nuclear state – rather than the resistances it provoked – we can gain a better understanding of the nuclear state's political logic and the discourses that speak to us through the now predominantly abandoned nuclear landscapes of the Cold War.

One question thus remains, namely whether there is a way of engaging with these ruins without aestheticising them, without falling into a Ratzelian ruin gaze. In a sense, this question is not unique to the study of the Cold War but has reverberated powerfully through a wide range of debates in human geography and beyond (DeSilvey & Edensor 2012). Bunkers were long seen as eliding the eye of the ruin gazer. As Paul Hirst argued in 2005, they 'fall into disuse and disrepair, but they don't really become ruins'. Instead, he explained, '[t]hey are, as blocks of mass concrete, still there, just as they were' (Hirst 2005: 214). Bunkers, in this view, are seen to produce rubble not ruins. And yet, today the heritage industry is increasingly converting nuclear bunkers into museums where the Cold War can be 'felt' and 'experienced' in often palatable and quaint ways. Whilst Hitler's last retreat in central Berlin was never allowed to become a site of memorialisation for concerns that it might prompt the 'wrong kind' of remembrance (Till 2005), this fear of the 'dangerous pull of nostalgia' does not seem to extend to Cold War bunkers (Beck 2011: 87). Whilst the story of some sites, such as the governmental nuclear bunker south of Bonn, is told by its museum through largely technical and apolitical storylines, the Berlin Underworld Society (*Berliner Unterwelten*), for instance, has tried to promote a more critical history of the Cold War.

Whilst Germany's Cold War bunkers are now turned into heritage sites with relatively high frequency, it is striking that missile storage sites have not (yet) been included in these attempts.[3] As we have seen, many of these places have either been completely dismantled (Figure 7.1), sunk into oblivion or been reused for a whole range of purposes. There may be a number of reasons for this. Firstly, these sites are often in rural locations rather than already established tourist locations. NATO, as we have seen in Chapter 5, went to great lengths to find such remote locations. Yet, there are also a whole range of other twentieth-century battlefields in relatively inaccessible locations which have nonetheless been converted into tourist sites. A second explanation, and perhaps a more convincing one, is that these ruins are very difficult to appropriate within existing narratives of the German Cold War, which tend to focus on the evils of the East German dictatorship on the one hand, and the superpower struggle on the other. More often than not, West Germany is presented as a passive pawn in a global game over which it

**Figure 7.1**  Abandoned missile camp in Bellersdorf (Hesse) 2015. Source: Author.

had little agency. There is, of course, a kernel of truth in this assessment, for the Federal Republic *was* a semi-sovereign state. The radically self-destructive nature of the tactical missile site, however, shines an interesting light on the relationship between the West German government and its NATO allies. It is difficult to argue that a state that was willing to expose itself to death at its own hands was *on the right side of history*, as it were, however much the eventual implosion of the East bloc may have proved Bonn's cold warriors right. Moreover, tactical missile sites continue to pose difficult questions about the continuing presence of nuclear weapons in Europe today. Is the Cold War really *over* and what is Germany's role in the ongoing conflict with Moscow?

It is thus possible to propose that there is something potentially disruptive in the ruins of the Cold War, an appreciation of which requires us to step outside of a romanticised conception of ruin value developed by Friedrich Ratzel and Albert Speer (a romanticisation that might reappear if the ruin was approached through an enchanted and affirmative mode of critique?). Speer, in his admiration of antique architecture, would have rejected the idea that there was anything at all aesthetic about the subterranean concrete of the nuclear bunker or indeed the missile silo. And yet, some ruins of the Cold War are today being approached in a Ratzelian way – as remnants of a destructive but necessary struggle for survival. Indeed, many former Cold War sites are visited by retired soldiers and military enthusiasts, many of whom are *nostalgic* for these places. A crucial way out of this aestheticisation of the ruin is provided by Walter Benjamin who read the ruin as an 'allegory' rather than a 'symbol', the latter remaining enthralled by the beauty of ruination (Benjamin 1977: 178). This is to reject the particular temporal aesthetics that Speer's theory of ruin value lends it, a kitsch gaze from the future onto the present that assumes that future generations would look at the present in much the same way as the present looked at the ruins of Roman antiquity (Stead 2003). For Benjamin, the ruin thus stands both for a mode of critique (ruination) and for the *frailty* of human life, a reminder that all human life is destined to end and be returned to the folds of nature.

Here, it is interesting to note a striking tension inherent in Cold War geopolitics, which the ruin brings to light. For Cold War geopolitics, as a discourse *about* space and a discourse *in* space, is marked by both a cyclical and an apocalyptic sense of time. On the one hand, Cold War geopoliticians, like their German geopolitical forefathers, would insist that world history was never more than the successive redistribution of power. No nation, not even a superpower, would ultimately be able to alter the iron laws of world politics, such as the struggle for space. On the other hand, however, there was a recognition amongst all but the most die-hard strategists that the arms race produced at least the possibility of an accidental Armageddon, sparked perhaps by a miscalculation or a technical failure. Although the bipolar distribution of power and the threat of mutually assured destruction were commonly described as producing an especially stable and durable world order for over four decades, nuclear geopolitics also expressed itself in the anxious and apocalyptic landscapes of missile silos, early warning systems, submarine pens and nuclear bunkers, the scope of which would suggest that *someone* at least was preparing for the apocalypse.

It is this second temporality, inherent in nuclear war, that exceeds the geopolitical logic of ruin value as articulated by Ratzel, Speer and others. Indeed, the implication of Benjamin's notion of the ruin, although he did not frame it in those terms, is the possibility of a world without humans, an idea which connects the anti-nuclear struggle of the Cold War era with contemporary resistance to anthropogenic climate change. In this, the ruins of the Cold War have the power to disrupt a cyclical view of history that remains at the heart of the modern geopolitical imagination.

## Endnotes

1   Note also that Latour himself has found intellectual inspiration in the geopolitician Carl Schmitt and his problematic friend–enemy distinction (Latour 2013). This highlights that the tradition of German geopolitics remains to this day a crucial reference point for scholars and writers with a wide range of political agendas. The vast majority of interest comes from those who have traced the renaissance of geopolitical ideas in post-Cold War Germany (Bach & Peters 2002; Bassin 2003; Behnke 2006; Ciută & Klinke 2010; Hoffmann 2012; Murphy & Johnson 2004; Reuber & Wolkersdorfer 2002).

2   In comparison to US civil defence, the West German programme was less nationalist in its rhetoric. US civil defence documents, translated by the German bureaucracy, would often make reference to the mythology of American nationalism (founding fathers, etc.) in ways that their German counterparts could not.

3   Note also the architectural violence with which former East German governmental buildings have been treated in recent years (Colomb 2007).

# References

I. Korps Stab (1966a) 'Kriegstagebuch 1211200Z', Military Archives Freiburg, BH 7–1/428.

I. Korps Stab (1966b) 'Nachrichten und Kommentare Fallex 66', Military Archives Freiburg 7–1/428.

III. Korps Chef des Stabes (1966) 'Inhalt von PSK Maßnahmen 181900Z', Military Archives Freiburg, BH 7–3/854.

III. Korps Fernschreibestelle (1966) 'Die Rede des Herrn Bundespräsidenten 201845Z', Military Archives Freiburg, BH 7–3/854

III. Korps Leitung (1966) 'G3 sofort verlegen !!! 222335Z', Military Archives Freiburg, BH 7–3/845.

III. Korps Leitungsstab (1966) 'FALLEX 66 Exercise *Top Gear* 180615Z', Military Archives Freiburg, BH 7–3/854.

Abrahamsson, C. (2013) 'On the genealogy of lebensraum', *Geographica Helvetica*, 68, pp. 37–44.

Adenauer K. (1946) '16. März 1946: Brief an William F. Sollmann, Pendle Hill, Wallingford/Pennsylvanien', available at http://www.konrad–adenauer.de/dokumente/briefe/brief-sollmann (accessed 5 January 2015).

Adenauer, K. (1949) '4. Dezember 1949: Artikel des Europakorrespondenten der amerikanischen Zeitung "Cleveland Plain Dealer", Leacacos, über ein Interview mit dem Bundeskanzler', available at http://www.konrad–adenauer.de/dokumente/interviews/cleveland–plain–dealers (accessed 5 January 2015).

Adenauer, K. (1951) '31. Januar 1951: Aufzeichnung des Bundeskanzlers zur Lage', available at http://www.konrad-adenauer.de/dokumente/aufzeichnungen/aufzeichnung-lage (accessed 5 January 2015).

Adenauer, K. (1954) 'Chancellor of Germany writes an introduction', *Life*, 10 May 1954.

Adenauer, K. (1957) 'Ansprache im Bayerischen Rundfunk zur Abrüstungspolitik', 3 July 1957, at http://www.konrad-adenauer.de/dokumente/reden/rundfunkansprache2 (accessed 5 January 2015).

*Cryptic Concrete: A Subterranean Journey Into Cold War Germany*, First Edition. Ian Klinke.
© 2018 John Wiley & Sons Ltd. Published 2018 by John Wiley & Sons Ltd.

Adey, P. (2008) 'Airports, mobility and the calculative architecture of affective control', *Geoforum*, 39(1), pp. 438–451.

Adey, P. and Anderson, B. (2012) 'Anticipating emergencies: Technologies of preparedness and the matter of security', *Security Dialogue*, 43, pp. 99–117.

Adey, P., Whitehead, M. and Williams, A. J. (2013) *From Above: War, Violence and Verticality*. London: Hurst.

Afheldt, H. (1983) *Defensive Verteidigung*. Hamburg: Rowolth.

Agamben, G. (1998) *Homo Sacer: Sovereign Power and Bare Life*. Stanford: Stanford University Press.

Agamben, G. (2005) *State of Exception*. Chicago: University of Chicago Press.

Agnew, J. (2003) *Geopolitics: Re-visioning World Politics*. London: Routledge.

Amoore, L. (2006) 'Biometric borders: Governing mobilities in the war on terror', *Political Geography*, 25(3), pp. 336–351.

Arkin, W. M. and Fieldhouse, R. W. (1986) *'Nuclear Battlefields': Der Atomwaffenreport*. Frankfurt: Athenäum.

Armitage, J. (2009) *Virilio: Selected Interviews*, London: Sage.

Atikinson, D. (2012) 'Encountering bare life in Italian Libya and colonial amnesia in Agamben', in Svirsky, M. and Bignall, S. (2012) *Agamben and Colonialism*, Edinburgh: University of Edinburgh Press, pp. 155–177.

Atkinson, D. and Cosgrove, D. (1998) 'Urban rhetoric and embodied identities: City, nation, and empire at the Vittorio Emanuele II Monument in Rome, 1870–1945', *Annals of the Association of American Geographers*, 88(1), pp. 28–49.

Atomwaffen a-z (2015) 'Alten-Buseck: ehemaliger Atomwaffenstandort, Deutschland', at http://www.atomwaffena-z.info/glossar/a/a-texte/artikel/8c8f32d201/alten-buseck.html (accessed 1 June 2015).

Bach, J. and Peters, S. (2002) 'The new spirit of German geopolitics', *Geopolitics*, 7(3), pp. 1–18.

Bachmann, V. (2009) 'From Jackboots to Birkenstocks: The civilianisation of German geopolitics in the twentieth century', *Tijdschrift voor Economische and Sociale Geographie*, 101(3), pp. 320–332.

Bald, D. (2005) *Die Bundeswehr: Eine kritische Geschichte 1955-2005*. München: C. H. Beck.

Bald, D. (2008) *Politik der Verantwortung*. Berlin: Aufbau.

Barnes, T. and Abrahamsson, C. (2015) 'Tangled complicities and moral struggles: The Haushofers, father and son, and the spaces of Nazi geopolitics', *Journal of Historical Geography*, 47, pp. 64–73.

Barnes, T. and Minca, C. (2013) 'Nazi Spatial Theory: The Dark Geographies of Carl Schmitt and Walter Christaller', *Annals of the Association of American Geographers*, 103(3), pp. 669–687.

Barry, A. (2013a) 'The translation zone: Between actor-network theory and international relations', *Millennium: Journal of International Studies*, 41(3), pp. 413–429.

Barry, A, (2013b) *Material Politics: Disputes Along the Pipeline*, Oxford: Wiley-Blackwell.

Bartolini, N. (2015) 'The politics of vibrant matter: Consistency, containment and the concrete of Mussolini's bunker', *Journal of Material Culture*, 20(2), pp. 191–210.

Bassin, M. (1987a) 'Race *contra* space: The conflict between German *geopolitik* and National Socialism', *Political Geography*, 6, pp. 115–134.

Bassin, M. (1987b) 'Imperialism and the nation state in Friedrich Ratzel's political geography', *Progress in Human Geography*, 11(4), pp. 473–495.

Bassin, M. (2003) 'Between realism and the "New Right": Geopolitics in Germany in the 1990s', *Transactions of the Institute of British Geographers*, 28, pp. 350–366.

Bauman, Z. (1989) *Modernity and the Holocaust*, Cambridge: Polity.

Bauamt Bonn der Bundesbaudirektion Berlin (1961) 'An das Bundesministerium des Innern z.Hd. von Herrn Oberst a.D. von Boelzig', 21 September 1961, Federal Archives Koblenz B106/50604.

Beck, J. (2011) 'Concrete ambivalence: Inside the bunker complex', *Cultural Politics*, 7(1), pp. 79–102.

Behnke, A. (2006) 'The Politics of geopolitik in post-Cold War Germany', *Geopolitics* 11(3), pp. 396–419.

Belcher, O., Martin, L., Secor, A. et al. (2008) 'Everywhere and nowhere: The exception and the topological challenge to Geography', *Antipode*, 40(4), pp. 499–503.

Bell, D. F. (2008) 'Bunker busting and bunker mentalities, or is it safe to be underground?', *South Atlantic Quarterly*, 107(2), pp. 213–229.

Benjamin, W. (1977) *The Origin of German Tragic Drama*. New York and London: Verso.

Bennett, L. (2011a) 'The Bunker: Metaphor, materiality and management', *Culture and Organization*, 17(2), pp. 155–173.

Bennett, L. (2011b) 'Bunkerology – a case study in the theory and practice of urban exploration', *Environment and Planning D*, 29, pp. 421–434.

Bennett, L. (2017) *In the Ruins of the Cold War Bunker: Affect, Materiality and Meaning Making*. London: Rowman and Littlefield International.

Berger Ziauddin, S. (2016) '(De)territorializing the home: The nuclear bomb shelter as a malleable site of passage', *Environment and Planning D*, Online first, DOI: https://doi.org/10.1177/0263775816677551

Berliner Zeitung (1966) '"Panzerfaust" gegen "revoltierenden Mob"', 18 November 1966.

Berliner Zeitung (1975) 'Überfall im Regierungsbunker: Minister Lebers Sekretärin niedergestochen', 19 March 1975.

Bernstein, J. M. (2004) 'Bare life, bearing witness: Auschwitz and the pornography of horror', *Parallax*, 10(1), pp. 2–16.

Beyer, E. (1960) 'Führungsprobleme im System der Landesverteidigung', *Ziviler Luftschutz*, 24(5), pp. 141–146.

Biess, F. (2009) '"Everybody has a chance": Nuclear angst, civil defence, and the history of emotions in postwar West Germany', *German History*, 27(2), pp. 214–243.

Billig, M. (1995) *Banal Nationalism*. London: Sage.

Blumentritt, G. (1952) *Deutsches Soldatentum im europäischen Rahmen*. Gießen: Verlag Westunion.

Blumentritt, G. (1956) *Die Rolle der Bundesrepublik im Rahmen der strategischen Abwehr der westlichen Welt*. Bad Godesberg: Politische Informationen.

Blumentritt, G. (1960) *Strategie und Taktik*. Konstanz: Akademische Verlagsgesellschaft Athenaion.

Boesler, K.-A. (1983) *Politische Geographie*. Stuttgart: Teubner.

Borch-Jacobsen, M. (1982) *The Freudian subject*. Houndsmills, Macmillan.

Borinski, P. (1989) 'Mitigating West Germany's strategic dilemmas', *Armed Forces and Society*, 15, pp. 531–549.

Bowman, I. (1942) 'Geography vs. geopolitics', *Geographical review*, 32(4), pp. 646–658.

Bratton, B. (2006) Foreword, in Virilio, P. (1977), *Speed and Politics*. Los Angeles: Semiotext(e), pp. 7–25.

Bregazzi, H. R. and Jackson, M. S. (2016) 'Agonism, critical political geography, and the new geographies of peace', *Progress in Human Geography*, available at http://journals. sagepub.com/doi/abs/10.1177/0309132516666687 (accessed 31 October 2017).

Breitman, R., Goda, N. J. W., Naftali, T. et al. (2005) *U.S. intelligence and the Nazis*, Cambridge: Cambridge University Press.

Bulletin (1954) 'Beitrag zur Rüstungsbeschränkung', 23 October 1954, in Kinkel, K. (1995), *Aussenpolitik der Bundesrepublik Deutschland: Dokumente von 1949-1994* (Bonn: Auswärtiges Amt), pp. 214–217.

Bundesamt für zivilen Bevölkerungsschutz (1961a) 'Jeder hat eine Chance', available at http://www.bbk.bund.de/SharedDocs/Downloads/BBK/DE/FIS/DownloadsDigitalisierte Medien/Broschueren/Jeder%20hat%20eine%20Chance.pdf?__blob=publicationFile (accessed 2 January 2017).

Bundesamt für zivilen Bevölkerungsschutz (1961b) 'Der Luftschutzhilfsdienst: Was er ist und was er will', available at http://www.bbk.bund.de/SharedDocs/Downloads/BBK/ DE/FIS/DownloadsDigitalisierteMedien/Broschueren/Luftschutzhilfsdienst.pdf?__ blob=publicationFile (accessed 2 January 2017).

Bundesamt für zivilen Bevölkerungsschutz (1965) 'Ausbildungsunterlagen für den Luftschutzdienst: Allgemeine Ausbildung', available at http://www.bbk.bund.de/ SharedDocs/Downloads/BBK/DE/FIS/DownloadsDigitalisierteMedien/Broschueren/ Ausbildungsunterlagen%20LSHD%201963.pdf?__blob=publicationFile (accessed 2 January 2017).

Bundesamt für zivilen Bevölkerungsschutz (1970) 'Das Bundesamt für Zivilen Bevölkerungsschutz', available at http://www.bbk.bund.de/SharedDocs/Downloads/ BBK/DE/FIS/DownloadsDigitalisierteMedien/Broschueren/BzB.pdf?__ blob=publicationFile (accessed 2 January 2017).

Bundesbaudirektion (1968a) 'An den Bundesschatzminister z.Hd. Herrn Ministerialrat Camerer o.V.i.A.', 26 July 1968, Federal Archives Koblenz B157/6094.

Bundesbaudirektion (1968b) 'An das Bundesschatzministerium z.Hd Herrn Min.-Rat Siemsen', 29 January 1968, Federal Archives Koblenz B157/6094.

Bundesbaudirektion (1968c) 'Erläuterungsbericht für das Bauvorhaben "Anlagen des THW, 2.Teil, "West", hier: Einbau einer CO-Warnanlage', Federal Archives Koblenz B157/6094.

Bundesfinanzministerium (1960) 'Erörterung eines Schutzraumbauprogramms der Bundesregierung im Bundesverteidigungsrat', 7 June 1960, Federal Archives Koblenz B 106/54720.

Bundesinnenministerium (1950) 'Sicherung der Bundesministerien gegen Luftangriffe', 9 December 1950, Federal Archives Koblenz B/106/201172.

Bundesinnenministerium (1955) 'Planung ziviler Notstandsmaßnahmen', 23 December 1955, Federal Archives Koblenz B/106/201172.

Bundesinnenministerium (1958a) 'Bericht über die Ergebnisse der Versuche mit deutschen Schutzbauten in den USA und über den Erfahrungsaustausch auf dem Gebiet des bau-lichen Luftschutzes', no date, Federal Archives Koblenz B 106/54720.

Bundesinnenministerium (1958b) 'In deiner Hand liegt der Schutz deiner Familie und deines Eigentums', no date, Federal Archives Koblenz B 106/17609.

Bundesinnenminsterium (1959) 'Gutachten zur Bestandsaufnahme für Anlagen des THW, Dr.-Ing. P. Walter', 15 April 1959, Federal Archives Koblenz, B157/6094.

Bundesinnenministerium (1960a) 'Arbeitsstäbe oberster Bundesbehörden in der Befehlsstelle der Bundesregierung', 1 April 1960, Federal Archives Koblenz B/106/201172.

Bundesinnenministerium (1960b) 'Grundannahmen für die Tätigkeit der Arbeitsstäbe in der Befehlsstelle der Bundesregierung', 1 April 1960, Federal Archives Koblenz B/106/201172.

Bundesinnenministerium (1960c) 'An den Herrn Bundesminister für Verkehr z.Hd. Herrn MinRat Busse o.V.i.A.', March 1960, Federal Archives Koblenz, B106/5041.

Bundesinnenministerium (1960d) 'Schutzraumbau', 4 July 1960, Federal Archives Koblenz B 106/54720.

Bundesinnenministerium (1961) 'An das Referat IV A 4 im Hause', 4 October 1961, Federal Archives Koblenz B106/50604.

Bundesinnenministerium (1963) 'Grundsätze für die taktische Beurteilung öffentlicher Schutzräume, 7 January 1963, Federal Archives Koblenz B 106/17609.

Bundesinnenministerium (1965) 'An das Bauamt Bonn der Bundesbaudirektion', 7 August 1965, Federal Archives Koblenz B157/6096.

Bundesinnenministerium (1966) 'An die Baudirektion Bauleitung Marienthal', 30 December 1966, Federal Archives Koblenz B157/6095.

Bundesinnenministerium (1969) 'Vermerk: Anlagen des THW in Marienthal; hier: Filteranlagen', 21 April 1969, Federal Archives Koblenz B157/6094.

Bundesinnenministerium (1972) 'Weißbuch zur zivilen Verteidigung der Bundesrepublik Deutschland', available at http://www.bbk.bund.de/SharedDocs/Downloads/BBK/DE/ FIS/DownloadsDigitalisierteMedien/Broschueren/Weißbuch%20zur%20zivilen%20 Verteidigung.pdf?__blob=publicationFile (accessed 2 January 2017).

Bundesinnenministerium (1973) 'Lichtbilder vom Bau der Anlage Marienthal', 20 December 1973, Federal Archives Koblenz B157/6095.

Bundesminister des Innern (1964) Zivilschutzfibel: Informationen, Hinweise, Ratschläge, Bad Godesberg: Bundesamt für zivilen Bevölkerungsschutz.

Bundesministerium der Verteidigung (1956) 'Vorschriften-Entwurf: Taktische Grundsätze des Heeres im Atomkrieg', (January 1956, Militärarchiv Freiburg BW 1/363052.

Bundesministerium der Verteidigung (1958) 'Vermerk: Gutachten über Kriegsmittel des Insdituts für Internationales Recht an der Universität Kiel', 15 April 1958, Militärarchiv Freiburg BW 1/313622.

Bundesministerium der Verteidigung (1959) 'Stellungnahme zu dem Aufsatz "Der Kampf mit gebrochenem Rückrat" von Lt Col. Robert B Rigg, Washington', 17 March 1959, Militärarchiv Freiburg BW 1/363052.

Bundesministerium der Verteidigung (1961a) 'Mehrstaatliche Vereinbarung über Sondermunitionslager', 2 September 1961, Militärarchiv Freiburg BW 1/59222/a.

Bundesministerium der Verteidigung (1961b) 'Second draft of Multi-national technical arrangement for the SAS support sites', 11 December 1961, Militärarchiv Freiburg BW 1/103529.

Bundesministerium der Verteiduging (1962) 'Munitionslager Lüdenscheid', 12 December 1961–4 June 1962, Militärarchiv Freiburg BW 1/97330.

Bundesministerium der Verteidigung (1963) 'Techn. Abkommen für den Betrieb und die Versorgung von SAS-Anlagen', 14 August 1963, Militärarchiv Freiburg BW 1/103524.

Bundesministerium der Verteidigung (1975a) 'Vereinbarkeit von US-Ausbildungsforderungen mit dem Gesetz über die Anwendung des unmittelbaren Zwanges', 6 June 1975, Militärarchiv Freiburg BW 1/106557.

Bundesministerium der Verteidigung (1975b) 'Besprechung mit Oberst i.G. Nickel, Fü L III 3, Oberstleutnant Kohler und Oberstleutnant Menzel, beide Fü L III 3', (day missing) September 1975, Militärarchiv Freiburg BW 1/106557.

Bundesministerium der Verteidigung (1981) 'Vortrag im Rahmen der Offizierweiterbildung: "Ist der Einsatz von A-Waffen mir dem Kriegsvölkerrecht (KVR) vereinbar?', 31 March 1981, Militärarchiv Freiburg BW 1/159166.

Bundesschatzminister (1968) 'Richtlinien für die Errichtung eines öffentlichen Schutzraumes in Verbindung mit einer unterirdischen Verkehrsanlage als Mehrzweckbau im Grundschutz in Berlin – Bauvorhaben "Excelsior"', January 1968, Federal Archives Koblenz B 106/56842.

Bundesschatzministerium (1967) 'An den Herrn Bundesminister für Verkehr', 21 September 1967, Federal Archives Koblenz B157/6095.

Bundestag (1954) 'Protokoll der 47. Sitzung', 7 October 1954, available at http://dipbt. bundestag.de/doc/btp/02/02047.pdf (accessed 6 December 2014).

Bundestag (1955a) 'Protokoll der 69. Sitzung', 24 February 1955, available at http:// dipbt.bundestag.de/doc/btp/02/02069.pdf (accessed 6 October 2014).

Bundestag (1955b) 'Protokoll der 100. Sitzung', 16th July 1955, available at http://dipbt. bundestag.de/doc/btp/02/02100.pdf (accessed 6 October 2014).

Bundestag (1958) 'Protokoll der 20. Sitzung', 22 March 1958, available at http://dipbt. bundestag.de/doc/btp/03/03020.pdf (accessed 6 October 2014).

Bundesverteidigungsministerium (1959) 'Grundsätzliche militärische Infrastrukturforderung für bauliche Schutzmaßnahmen in ständigen Truppenunterkünften', 8 October 1959, Federal Archives Koblenz B 106/54720.

Buteux, P. (1983) The Politics of Nuclear Deterrence in NATO 1965-1980. Cambridge: Cambridge University Press.

Campbell, D. (2005) 'The biopolitics of security: Oil, empire and the Sports Utility Vehicle', American Quarterly, 57(3), pp. 943–972.

Carter, S., Kirby, P. and Woodyer, T. (2015) 'Ludic – or playful – geopolitics', in M. Benwell and P. Hopkins (eds), Children, Young People and Critical Geopolitics. Farnham: Ashgate.

Carter-White, R. (2013) 'Towards a spatial historiography of the Holocaust: Resistance, film, and the prisoner uprising at Sobibor death camp', Political Geography, 33, pp. 21–30.

Christaller, W. (1955) 'Subkontinente', Zeitschrift für Geopolitik, 26(10), pp. 603–614.

Cioc, M. (1988) Pax atomica: The nuclear defense debate in West Germany during the Adenauer era. New York: Columbia University Press.

Ciută, F. (2015) 'Call of duty: Playing video games with IR', Millennium: Journal of International Studies, 44(2), pp. 197–215.

Ciută, F. and Klinke, I. (2010) 'Lost in conceptualization: Reading the "new Cold War" with critical geopolitics', Political Geography, 29(6), pp. 323–332.

Clark, N. (2013a) 'Geopolitics at the threshold', Political Geography, 37, pp. 48–50.

Clark, N. (2013b) 'Geoengineering and geologic politics', *Environment and Planning A*, 45(12), pp. 2825–2832.

Clausewitz, C. v. (1832[1984]) *On War*. Princeton: Princeton University Press.

Coaffee, J. (2006) 'From counter-terrorism to resilience', *The European Legacy*, 11(4), pp. 389–403.

Coaffee, J., O'Hare, P. and Hawkesworth, M. (2009) 'The visibility of (in)security: The aesthetics of planning urban defences against terrorism', *Security Dialogue*, 40(4–5), pp. 489–511.

Cocroft, W., Thomas, R. J. C., and Barnwell, P. S. (2005) *Cold War: Building for Nuclear Confrontation 1946–89*. London: English Heritage.

Cohen, S. B. (1975) *Geography and Politics in a World Divided*. Oxford: Oxford University Press.

Cohn, C. (1987) 'Sex, death and the rational world of defence intellectuals', *Journal of Women in Culture and Society*, 12, pp. 687–718.

Coleman, M. (2011) 'Colonial war: Carl Schmitt's deterritorialization of enmity', in S. Legg (ed.) *Spatiality, sovereignty and Carl Schmitt: Geographies of the nomos* (London: Routledge), pp. 127–142.

Collier, S. and Lakoff, A. (2015) 'Vital systems security: Reflexive biopolitics and the government of emergency', *Security Dialogue*, 32(2), pp. 19–51.

Colomb, C. (2007) 'Requiem for a lost Palast: "revanchist urban planning" and "burdened landscapes" of the German Democratic Republic in the new Berlin', *Planning Perspectives* 22(3), pp. 283–323.

Cowen, D. and Gilbert, E. (2007) *War, Citizenship, Territory*. London: Routledge.

Dählmann, H. (1953) 'Gedanken über den Schutzraumbau', *Ziviler Luftschutz*, 17(7/8), pp. 169–171.

Daitz, W. (1943) *Lebensraum und gerechte Weltordnung: Grundlagen einer Anti-Atlantikcharta*. Amsterdam: De Amsterdamsche Keurkamer.

Dalby, S. (1988) 'The Soviet Union as Other', *Alternatives*, 13, pp. 415–442.

Dalby, S. (1990a) 'American security discourse and the persistence of geopolitics', *Political Geography Quarterly*, 9, pp. 171–188.

Dalby, S. (1990b) *Creating the Second Cold War: The Discourse of Politics*. London: Pinter.

Dalby, S. (2007) 'Anthropocene geopolitics: Globalisation, empire, environment and critique', *Geography Compass*, 1, pp. 103–18.

Danielsson, S. K. (2009) 'Creating genocidal space: Geographers and the discourse of annihilation, 1880–1933', *Space and Polity*, 13, pp. 55–68.

Davis, S. (2008) 'Military landscapes and secret science: The case of Orford Ness', *Cultural Geographies*, 15, pp. 143–149.

Davis, T. C. (2007) *Stages of Emergency: Cold War Nuclear Civil Eefence*. Durham and London: Duke University Press.

Dehio, L. (1955) *Deutschland und die Weltpolitik im 20. Jahrhundert*. Frankfurt am Main and Hamburg: Fischer.

Der Derian, J. (2001) *Virtuous War: Mapping the Military-Industrial-Media-Entertainment Network*. Boulder: Westview.

Der Spiegel (1951) 'So nicht', 3/51, 4.

Der Spiegel (1955a) 'Was sag' ich meinem Sohn?' 9/14, 7–12.

Der Spiegel (1955b) 'Überholt von Pfeil und Bogen', 9/29, 7–10.

Der Spiegel (1957a) 'Die Bombe im Schiff', 11/20, 12.

Der Spiegel (1957b) 'Raketenbewaffnung: Vom Schild zum Schwert', 11/46, 13–14.

Der Spiegel (1962) 'Bedingt abwehrbereit', 16/41, 34–53.

Der Spiegel (1964) 'Verstrahlte Erde', 18/52, 16.

Der Spiegel (1966a) 'Fallex 66: Tischtuch für Generale', 20/44, 30.

Der Spiegel (1966b) 'Bier im Berg', 20/43, 27–28.

Der Spiegel (1967) 'Zusammenbruch am vierten Tag?', 21/32, 26–27.

Der Spiegel (1983) 'Kinder des Lichts, Kinder der Finsternis', 37/42, 30–32.

Der Spiegel (1984) 'Sieben Eide', 38/16, 66–76.

Der Spiegel (1985a) 'Armee gegen Streikende', 39/27, 15.

Der Spiegel (1985b) 'Auf breiter Front', 39/26, 32–33.

Der Spiegel (2013) 'USA Machen Alt-Atombombe zu Allzweckwaffe', available at http://www.spiegel.de/wissenschaft/technik/b61-atombombe-modernisierung-umfangreicher-als-bekannt-a-931642.html (accessed 30 January 2017).

DeSilvey, C. and Edensor, T. (2012) 'Reckoning with ruins', *Progress in Human Geography*, 37(4), pp. 465–485.

Deutsche Gesellschaft für Auswärtige Politik (1969) *Mittlere Mächte in der Weltpolitik*. Opladen: Leske Verlag.

Deutsche Societät Beratender Ingenieure (1961) 'THW Marienthal: Erläuterungsbericht/Kostenvoranschlag', 15 December 1962, Federal Archives Koblenz B106 201170.

Die Welt (2015) 'Amerikaner will Luxus Bunker in Thüringen bauen', available at https://www.welt.de/finanzen/immobilien/article143375523/Amerikaner-will-Luxus-Bunker-in-Thueringen-bauen.html (accessed 30 January 2017).

Die Zeit (1966a) 'Regierung unter Tage', 21 October 1966 43, 7.

Die Zeit (1966b) 'Der Aufstand der Linken', 2 December 1968 49, 8.

Die Zeit (2016) 'Atombombe? Nein danke!', available at http://www.zeit.de/2017/01/nukleare-abruestung-deutschland-atombombe-atomwaffen (accessed 12 February 2017).

Diester, J. (2009) *Geheimakte Regierungsbunker: Tagebuch eines Staatsgeheimnisses*, Düsseldorf: Verlagsanstalt Handwerk GmbH.

Diester, J. and Karle, M. (2013) *Plan B: Bonn, Berlin und ihre Regierungsbunker*. Düsseldorf: Verlagsanstalt Handwerk GmbH.

Diken, B. and Laustsen, C. B. (2004) *The Culture of Exception: Sociology Facing the Camp*. London, UK: Routledge.

Diken, B. and Laustsen, C. B. (2006) 'The camp', *Geografiska Annaler*, 88B, 4, pp. 443–552.

Dillon, M. and Lobo-Guerrero, L. E. (2008) 'Biopolitics of security in the 21st century: An introduction', *Review of International Studies*, 34(2), pp. 265–292.

Dittmer, J. (2013) 'Geopolitical assemblage and complexity', *Progress in Human Geography*, 38(3), pp. 385–401.

Dittmer, J. and Gray, N. (2010) 'Popular geopolitics 2.0: towards new methodologies of the everyday', *Geography Compass*, 4(11), pp. 1664–1677.

Dittmer, J. and Sharp, J. (2014) *Geopolitics: An Introductory Reader*. London: Routledge.

Dixon, D. (2014) 'The way of the flesh: Life, geopolitics and the weight of the future', *Gender, Place and Culture*, 21(2), pp. 136–151.

Dodds, K. (2003) 'Cold War geopolitics', in J. Agnew, K. Mitchell and G. Toal (eds), *Companion to Political Geography*. Malden: Blackwell, pp. 204–218.

Dodds, K. (2005) *Global Geopolitics: A Critical Introduction*. London: Routledge.

Dodds, K. (2007) *Geopolitics: A Very Short Introduction*. Oxford: Oxford University Press.

Dorn, W. (2002) *So heiss war der kalte Krieg: Fallex 66*. Berlin: Dittrich Verlag.

Dowler, L. and Sharp, J. (2001) 'A feminist geopolitics?', *Space and Polity*, 5(3), pp. 165–176.

Driver, F. and Gilbert, D. (1998) 'Heart of empire? Landscape, space and performance in imperial London', *Environment and Planning D*, 16(1), pp. 11–28.

Duffield, M. (2011) 'Total war as environmental terror: linking liberalism, resilience and the bunker', *South Atlantic Quarterly*, 110(3), pp. 757–769.

Dumas, L. J. (1980) 'Human fallibility and weapons', *Bulletin of Atomic Scientists*, 36(9), pp. 15–20.

Dwinger, E. E. (1957) *Es geschah im Jahre 1965*. Salzburg and München: Pilgrim.

Ebeling, W. (1986) *Schlachtfeld Deutschland?* Friedberg: Podzun-Pallas-Verlag.

Edkins, J. (2000) 'Sovereign power, zones of indistinction, and the camp', *Alternatives: Global, Local, Political*, 25(3), pp. 3–25.

Elden, S. (2009) *Terror and Territory: The Spatial Extent of Sovereignty*. Minneapolis: University of Minnesota Press.

Elden, S. (2010) 'Reading Schmitt geopolitically', *Radical Philosophy*, 161, pp. 18–26.

Elden, S. (2013) 'Secure the volume: Vertical geopolitics and the depth of power', *Political Geography*, 34, pp. 35–51.

Esposito, R. (2008) *Bios: Biopolitics and Philosophy*. Minneapolis: University of Minnesota Press.

Farish, M. (2004) 'Another anxious urbanism: Simulating defence and disaster in Cold War America', in S. Graham (ed.), *Cities, War, and Terrorism: Towards an Urban Geopolitics*. Oxford: Blackwell, pp. 93–109.

Farish, M. (2010) *The Contours of America's Cold War*. Minneapolis: University of Minnesota Press.

Farish, M. and Monteyne, D. (2015) 'Introduction: Histories of Cold War cities', *Urban History*, 42(4), pp. 543–546.

Featherstone, D. (2012) *Solidarity: Hidden Histories and Geographies of Internationalism*. John Wiley & Sons, Ltd.

Featherstone, M. (2005) 'Ruin value', *Journal for Cultural Research*, 9(3), pp. 301–320.

Fest, J. (2005) *Inside Hitler's Bunker: The Last Days of the Third Reich*. London: Pan Books.

Flint, C. (2005) *The Geography of War and Peace: From Death Camps to Diplomats*. Oxford: Oxford University Press.

Flockhardt, T. (2013) 'NATO's nuclear addiction – 12 steps to "kick the habit"', *European Security*, 22(3), pp. 271–287.

Forbes (2015) 'Billionaire bunkers: Exclusive look inside the world's largest planned doomsday escape', available at http://www.forbes.com/sites/jimdobson/2015/06/12/billionaire-bunkers-exclusive-look-inside-the-worlds-largest-planned-doomsday-escape/#3d0a715b5afb (accessed 20 January 2017).

Forty, A. (2012) *Concrete and Culture: A Material History*. London: Reaktion.

Foucault, M. (1976) *Society Must Be Defended: Lectures at the Collège de France, 1975–1976*. New York: Picador.

Foucault, M. (1978) *The History of Sexuality Part I: The Will to Knowledge*. London: Penguin.

Freedman, L. (2015) *Strategy: A History*. Oxford: Oxford University Press.

Fregonese, S. (2012a) 'Urban geopolitics 8 years on: Hybrid sovereignties, the everyday and geographies of peace', *Geography Compass*, 6, 5, 290–303.

Fregonese, S. (2012b) *War and the City: Urban Geopolitics in Lebanon*, London: I.B. Tauris.

Fregonese, S. and Ramadan, A. (2015) 'Hotel geopolitics: A research agenda', *Geopolitics*, 20, pp. 793–813.

Freitag, J. and Hensel, H. (2010) *Honeckers geheimer Bunker 5001*. Stuttgart: Motorbuchverlag.

Freud, S. (1913[2001]) *Totem and Taboo*. London: Routledge.

Freud, S. (1920[2001]) *Beyond the Pleasure Principle, Group Psychology and Other Works*. London: Vintage.

Freud, S. (1930[1989]) *Civilization and its Discontents*. London: Norton.

Friedrichs, J. H. (2008) 'Massenunterkunft, Atombunker, Kunstobjekt: Bunkeranlagen im Nachkriegsdeutschland', in I. Marszolek and M. Buggeln (eds), *Bunker: Kriegsort, Zuflucht, Erinnerungsraum*. Frankfurt and New York: Campus, pp. 245–260.

Fuhrmeister, C. and Mittig, H. E. (2008) 'Albert Speer und die "Theorie vom Ruinenwert" (1969) – der lange Schatten einer Legende', in I. Marszolek and M. Buggeln (eds), *Bunker: Kriegsort, Zuflucht, Erinnerungsraum*. Frankfurt and New York: Campus, pp. 225–243.

Gallaher, C., Dahlman, C. T., Gilmartin, M. et al. (2009) *Key Concepts in Political Geography*. London: Sage.

Gane, M. (1999) 'Paul Virilio's bunker theorizing', *Theory, Culture and Society*, 16(5/6), pp. 85–102.

Garrett, B. L. (2011) 'Shallow excavation, a response to Bunkerology', *Environment and Planning D*, available at http://societyandspace.org/2011/06/10/shallow-excavation-a-response-to-bunkerology-bradley-l-garrett (accessed 31 October 2017).

Geenen, E. (2006) 'Der kalkulierte Tod, in Bundesbauamt für Bauwesen und Raumordnung', Bundesamt für Bauwesen und Raumordnung/Stiftung Haus der Geschichte der Bundesrepublik Deutschland (eds.), *Der Regierungsbunker*. Berlin und Tübingen: Ernst Wasmuth Verlag, pp. 42–51.

Geist, E. (2012) 'Was there a real "mineshaft gap"? Bomb shelters in the USSR, 1945–1962', *Journal of Cold War Studies*, 14(2), pp. 3–28.

Gemeindeverwaltung Dernau (1971) 'An die Bundesbaudirektion', 23 July 1971, Federal Archives Koblenz B157/6094.

Giaccaria, P. and Minca, C. (2011a) 'Topographies/topologies of the camp: Auschwitz as a spatial threshold', *Political Geography*, 30, pp. 3–12.

Giaccaria, P. and Minca, C. (2011b) 'Nazi biopolitics and the dark geographies of the selva', *Journal of Genocide Research*, 13(1–2), pp. 67–84.

Giaccaria, P. and Minca, C. (2016) 'Life in space, space in life: Nazi topographies, geographical imaginations, and *Lebensraum*', *Holocaust Studies: A journal of Culture and History*, 22(2–3), pp. 151–171.

Giles, I. (1990) 'Introduction', in A. Watkins (ed.), *Footnotes to History: Selected Speeches and Writings of Edmund A. Walsh S. R.* Washington: Georgetown University Press.

Graham, S. (2004a) 'Vertical geopolitics: Baghdad and after', *Antipode*, 36(1), pp. 12–23.

Graham, S. (2004b) *Cities, War, and Terrorism: Towards an Urban Geopolitics*. Oxford: Blackwell.

Graham, S. (2006) 'Cities and the "War on Terror"', *Journal of Urban and Regional Research*, 30, pp. 255–276.

Graham, S. (2009) 'Cities as battlespace', *City*, 13(4), pp. 383–402.

Graham, S. (2010) *Cities Under Siege: The New Military Urbanism*. London: Verso.

Graham, S. (2016) *Vertical: The City from Satellites to Bunkers*. London: Verso.

Greenwood, T. (1952) 'Albrecht Haushofers letztes Werk', *Zeitschrift für Geopolitik*, 23(4), pp. 241–254.

Gregory, D. (2004) *The Colonial Present: Afghanistan, Palestine, Iraq*. Oxford: Blackwell.

Gregory, D. (2006) 'The black flag: Guantanamo bay and the space of exception', *Geografiska Annaler*, 88B(4), pp. 405–427.

Gregory, D. (2007) 'Vanishing points', in D. Gregory and A. Pred (eds), *Violent Geographies: Fear, Terror and Political Violence*. London: Routledge, pp. 205–236.

Gregory, D. (2010) 'Seeing red: Baghdad and the event-ful city', *Political Geography*, 29, pp. 266–279.

Gregory, D. (2010) 'War and peace', *Transactions of the Institute of British Geographers*, 35, pp. 154–186.

Gregory, D. (2011) '"Doors into nowhere": Dead cities and the natural history of destruction', in P. Meusburger, M. Heffernan and E. Wunder (eds), *Cultural Memories: The Geographical Point of View*. Springer: Heidelberg, London and New York, pp. 249–283.

Gregory, D. and Pred, A. (2007) *Violent Geographies: Fear, Terror and Political Violence*. London: Routledge.

Gross, G. P. (2016) *The Myth and Reality of German Warfare: Operational Thinking from Moltke to Heusinger*. Lexington: University Press of Kentucky.

Gückelhorn, W. (2002) *Lager Rebstock*. Aachen: Helios.

Guderian, H. (1937[1992]) *Achtung-Panzer! The Development of Tank Warfare*. London: Cassell.

Guderian, H. (1950) *Kann Westeuropa verteidigt werden?* Göttingen: IM Plesse.

Guderian, H. (1951a) *So geht es nicht! Ein Beitrag zur Frage der Haltung Westdeutschlands*. Göttingen: IM Plesse.

Guderian, H. (1951b) 'Raum und Zeit in der modernen Kriegsführung', *Zeitschrift für Geopolitik*, 22(1), pp. 7–13.

Hagen, J. and Ostergren, R. (2006) 'Spectacle, architecture and place at the Nuremberg Party Rallies: Projecting a Nazi vision of past, present and future', *Cultural Geographies*, 13, pp. 157–181.

Halas, M. (2014) 'Searching for the perfect footnote: Friedrich Ratzel and the others at the roots of lebensraum', *Geopolitics*, 19(1), pp. 1–18.

Hampe, E. (1952) 'Aufbau des zivilen Luftschutzes', *Ziviler Luftschutz*, 16(1), pp. 5–7.

Hampe, E. (1956a) *Im Spannungsfeld der Luftmächte: Eine Einführung in Luftgefahr und Bevölkerungsschutz*. Köln: Maximilian-Verlag Köln.

Hampe, E. (1956b) *Strategie der zivilen Verteidigung: Studie zu einer brennenden Zeitfrage*. Frankfurt a. M.: R. Eisenschmidt Verlag.

Handel, A. (2014) 'Gated/gating community: The settlement complex in the West Bank', *Transactions of the Institute of British Geographers*, 39, pp. 504–517.

Haney, D. (forthcoming) *Blood, Soil, Building: The Nazi Cultural Landscape* (under review).

Hardt, M. and Negri, A. (2000) *Empire*. Cambridge: Harvard University Press.

Haushofer, A. (1951) *Allgemeine Politische Geographie und Geopolitik*. Heidelberg: Kurt Vowinckel Verlag.

Haushofer, K. (1925) 'Politische Erdkunde und Geopolitik', in E. Drygalski (ed.), *Freie Wege vergleichender Erdkunde*. München und Berlin: Druck und Verlag von R. Oldenbourg, pp. 86–103.

Haushofer, K. (1926[1979]) 'Vergleich des Lebens-Raumes Deutschlands mit dem seiner Nachbarn unter besonderer Berücksichtigung der wehrgeographischen Lage der Vergleichs-Staaten', unpublished manuscript, in H.-A. Jacobsen (ed.), *Karl Haushofer: Leben und Werk Band I*. Boppard: Harald Boldt Verlag, pp. 524–536.

Haushofer, K. (1928) 'Rheinische Geopolitik', in K. Haushofer (ed.), *Der Rhein: Sein Lebensraum/sein Schicksal (1*. Band/Erdraum und Erdkräfte*)*. Berlin: Kurt Vowinckel Verlag, pp. 1–16.

Haushofer, K. (1934a) *Weltpolitik von heute*. Berlin: Zeitgeschichte.

Haushofer, K. (1934b) 'Geopolitische Grundlagen', in H.-A. Jacobsen (ed.), *Karl Haushofer: Leben und Werk Band I*. Boppard: Harald Boldt Verlag, pp. 558–606.

Haushofer, K. (1935) 'Die raumpolitischen Grundlagen der Weltgeschichte', in K. von Müller and P. Rohden (eds), *Knaurs Weltgeschichte: Von der Urzeit zur Gegenwart*. Berlin: Verlag von Th. Knaur Nachf, pp. 11–43.

Haushofer, K. (1944) 'Nostris ex ossibus. Gedanken eines Optimisten', unpublished, in H.-A. Jacobsen (ed.), *Karl Haushofer: Leben und Werk Band I*. Boppard: Harald Boldt Verlag, pp. 634–639.

Haushofer, K. (1946) 'Erklärung an unseren Sohn Heinz und unseren Familien-Anwalt Dr. Carl Beisler', in H.-A. Jacobsen (ed.), *Karl Haushofer: Leben und Werk Band I*. Boppard: Harald Boldt Verlag, pp. 447.

Haushofer, K. (1955) 'Die Inselwelt des Mittelmeers zwischen Asien und Australien', *Zeitschrift für Geopolitik*, 26(10), pp. 233–237.

Headquarters Northern Army Group (1961) 'A multi-national technical arrangement between the designated representatives of Canada, the Federal Republic of Germany, the Kingdom of Belgium, the Kingdom of the Netherlands, the United Kingdom of Great Britain and Northern Ireland and the United States of America', 27 October 1961, Militärarchiv Freiburg BW 1/59222/a.

Heefner, G. (2012) *The Missile Next Door: The Minuteman in the American Heartland*. Cambridge and London: Harvard University Press.

Hell, J. (2009) 'Katechon: Carl Schmitt's imperial theology and the ruins of the future', *The Germanic Review: Literature, Culture, Theory*, 84(4), pp. 283–326.

Hell, J. and Steinmetz, G. (2014) 'Ruinopolis: Post-imperial theory and learning from Las Vegas', *International Journal of Urban and Regional Research*, 38(3), pp. 1047–1068.

Hennig, R. (1935) *Einführung in die Geopolitik*. Leipzig and Berlin: B. G. Teubner.

Hepple, L. (2009) 'Dudley Stamp and the *Zeitschrift für Geopolitik*', *Geopolitics*, 13(2), pp. 386–395.

Herb, G. (2005) 'The geography of peace movements', in C. Flint (ed) *The Geography of War and Peace: From Death Camps to Diplomats*. Oxford: Oxford University Press, pp. 347–368.

Herwig, H. (1999) 'Geopolitik: Haushofer, Hitler and lebensraum', *Journal of Strategic Studies*, 22(2–3), pp. 218–241.

Heske, H. (1986) 'Political geographers of the past III: German geographical research in the Nazi period: a content analysis of the major geography journals, 1925-1945', *Political Geography Quarterly*, 5(3), pp. 367–281.

Heske, H. (1987) 'Karl Haushofer: His role in German geopolitics and in Nazi politics', *Political Geography Quarterly*, 6(2), pp. 135–144.

Heyden, G. (1958) *Kritik der deutschen Geopolitik: Wesen und soziale Funktion einer reaktionären soziologischen Studie*. Berlin: Dietz Verlag.

Hirst, P. (2005) *Space and Power: Politics, War and Architecture*. London: Polity.

Hitler, A. (1925[1939]) *Mein Kampf*. London, New York and Melbourne: Hurst and Blackett.

Hitler, A. (1928[2006]) *Hitler's Second Book: The Unpublished Sequel to Mein Kampf*. New York: Enigma.

Hitler, A. (1945) 'Hitler's "Scorched Earth" Decree (Nero Decree)', March 19, 1945, in United States Chief Counsel for the Prosecution of Axis Criminality, *Nazi Conspiracy and Aggression*. Supplement B. Washington, DC: United States Government Printing Office, 1948. Speer Document 27, pp. 950–51. English translation edited by GHI staff, available at http://germanhistorydocs.ghi-dc.org/sub_document.cfm?document_id=1590 (accessed 11 February 2014).

Hochtief (2014) 'My company, our history', available at http://www.hochtief.com/hochtief_en/74.jhtml;jsessionid=DCF768CD241FF1B4E6CACD4FEF9BD48D (accessed 11 February 2014).

Hoffmann, N. (2012) *Renaissance der Geopolitik: Die deutsche Sicherheitspolitik nach dem Kalten Krieg*, Wiesbaden: Springer Verlag für Sozialwissenschaften.

Holmes, A. (2013) *Social Unrest and American Military Bases in Turkey and Germany Since 1945*. Cambridge: Cambridge University Press.

Hoppe, C. (1992) *Zwischen Teilhabe und Mitsprache: Die Nuklearfrage in der Allianzpolitik Deutschlands 1959-1966*. Baden-Baden: Nomos.

Hornblum, A. M., Newman, J. L. and Dober, G. J. (2013) *Against Their Will: The Secret History of Medical Experimentation on Children in Cold War America*. Basingstoke: Palgrave.

Housden, M. (2003) *Hans Frank, Lebensraum and the Holocaust*. Basingstoke: Palgrave.

Hughes, R. (2010) 'Gameworld geopolitics and the genre of the quest', in F. MacDonald, R. Hughes and K. Dodds (eds), *Observant States: Geopolitics and Visual Culture* (London: I. B. Tauris), pp. 123–142.

Informationsbüro für Friedenspolitik (1982) *Lagerung und Transport von Atomwaffen*. München: Informationsbüro für Friedenspolitik.

Ingram A. and Dodds, K. (2008) *Spaces of Security and Insecurity: Geographies of the War on Terror*. Aldershot: Ashgate.

Ingram, A. (2001) 'Alexander Dugin: Geopolitics and neo-fascism in post-Soviet Russia', *Political Geography*, 20, pp. 1029–1051.

Jacobsen, K. (2013) 'Why Freud matters: Psychoanalysis and International Relations revisited', *International Relations*, 27, pp. 393–416.

Jagemann, E. (1955) *Die raumpolitischen Grundlagen Europas: Gesehen von einem Geopolitiker aus der Schule Albrecht Haushofers*. Wolfshagen-Scharbeutz: Franz Westphal.

Janta, L. Rieck, H. and Riemenschneider, M. (1989) *Kreis Ahrweiler unter dem Hakenkreuz*. Bad Neuenahr-Ahrweiler: Landkreis Ahrweiler.

Jones, L. and Sage, D. (2010) 'New directions in critical geopolitics: An introduction', *GeoJournal*, 75(4), pp. 315–325.

Jones, M., Jones, R. and Woods, M. (2014) *An Introduction to Political Geography: Space, Place and Politics*. London: Routledge.

Jordan, P. (1954) *Atomkraft: Drohung und Versprechen*. München: Wilhelm Heyne Verlag.

Jordan, P. (1957) *Der gescheiterte Aufstand: Betrachtungen zur Gegenwart*. Frankfurt: Vittorio Klostermann.

Jungbluth, U. (2000) *Wunderwaffen im KZ "Rebstock"*. Briedel: Rhein-Mosel Verlag.

Jureit, U. (2012) *Das Ordnen von Räumen: Territorium und Lebensraum im 19. und 20. Jahrhundert*. Hamburg: Hamburger Edition.

Kaplan, R. D. (2012) *The Revenge of Geography: What the Map Tells Us About Coming Conflicts and the Battle Against Fate*. New York: Random House.

Kearns, G. (2009) *Geopolitics and Empire: The Legacy of Halford Mackinder*. Oxford: Oxford University Press.

Kearns, G. (2010) 'Geography, geopolitics and empire', *Transactions of the Institute of British Geographers*, 35(2), pp. 187–203.

Kearns, G. (2011) 'Echoes of Carl Schmitt among the ideologists of the new American Empire', in S. Legg (ed.), *Spatiality, Sovereignty and Carl Schmitt: Geographies of the Nomos*. London: Routledge, pp. 74–90.

Kennan, G. F. (Mr X) (1947) 'The sources of Soviet conduct', *Foreign Affairs*, 25(4), pp. 566–582.

Kingsbury, P. (2008) 'Did somebody say jouissance? On Slavoj Žižek, consumption, and nationalism', *Emotion, Space and Society*, 1, 48–55.

Kingsbury, P. (2010) 'Locating the melody of the drives', *The Professional Geographer*, 62(4), pp. 519–533.

Kingsbury, P. and Pile, S. (2014) 'The unconscious, transference, drives and other things tied to geography', in P. Kingsbury and S. Pile (eds), *Psychoanalytic Geographies*. Farnham: Ashgate, pp. 1–39.

Kissinger, H. A. (1957[1969]) *Nuclear Weapons and Foreign Policy*. Toronto: W. W. Norton and Company.

Kjellén, R. (1917) *Der Staat als Lebensform*. Leipzig: S. Hirzel.

Kjellén, R. (1920) *Grundriss zu einem System der Politik*. Leipzig: S. Hirzel Verlag.

Klingmüller, A. (1955) 'Luftschutz und unterirdische Verkehrswege', *Ziviler Luftschutz*, 19(9), pp. 199–200.

Klinke, I. (2011) 'Geopolitics in Germany: *Return* of the living dead?', *Geopolitics*, 16(3), pp. 707–726.

Klinke, I. (2013) 'Chronopolitics: A conceptual matrix', *Progress in Human Geography* 37(5): pp. 673–690

Klinke, I. (2015) 'The bunker and the camp: Inside West Germany's nuclear tomb', *Environment and Planning D*, 33(1), pp. 154–168.

Klinke, I. (2016) 'Self-annihilation, nuclear play, and West Germany's compulsion to repeat', *Transactions of the Institute of British Geographers*, 41(1), pp. 109–120.

Klinke, I. and Perombelon, B. (2015) 'Notes on the desecuritisation of the Rhineland frontier', *Geopolitics*, 20(4), pp. 836–852.

Koopman, S. (2011) 'Let's take peace to pieces', *Political Geography*, 30, pp. 193–194.

Kost, K. (1988) *Die Einflüsse der Geopolitik auf die Forschung und Theorie der Politischen Geographie von ihren Anfängen bis 1945*. Bonn: Ferd. Dümmlers.

Kost, K. (1989) 'The conception of politics in political geography and geopolitics in Germany until 1945', *Political Geography Quarterly*, 8(4), pp. 369–384.

Kremer, B. (1963) *Der kluge Mann baut tief*. München: Osang Verlag.

Kremer, B. (1966) *Die Kunst zu Überleben: Zivilverteidigung in der Bundesrepublik*. München: Osang Verlag.

Krier, L. (1985[2013]) *Albert Speer: Architecture 1932-1942*. New York: Monacelli.

Kuklick, B. (2006) *Blind Oracles: Intellectuals and War from Kennan to Kissinger*. Princeton: Princeton University Press.

Kundnani, H. (2014) *The Paradox of German Power*. London: Hurst and Company.

Kuus, M. (2014) *Geopolitics and Expertise: Knowledge and Authority in European Diplomacy*. Malden and Oxford: Wiley Blackwell.

Laclau, E. (2007) 'Bare life or social indeterminacy', in M. Calarco and S. DeCaroli (eds), *Giorgio Agamben: Sovereignty and Life*. Stanford: Stanford University Press, pp. 11–22.

Laruelle, M. (2004) 'The two faces of contemporary Eurasianism: An imperial version of Russian nationalism', *Nationalities Papers*, 32(1), pp. 115–136.

Latour, B. (2004) 'Why has critique run out of steam? From matters of fact to matters of concern', *Critical inquiry*, 30(2), pp. 225–248.

Latour, B. (2013) 'War and peace in an age of ecological conflicts', available at http://www.bruno-latour.fr/sites/default/files/130-VANCOUVER-RJE-14pdf.pdf (accessed 9 January 2017).

Legg, S. (2011) *Spatiality, Sovereignty and Carl Schmitt: Geographies of the Nomos*. London: Routledge.

Lemke, B. (2007) 'Zivile Kriegsvorbereitungen in unterschiedlichen Staats- und Gesellschaftssystemen: Der Luftschutz im 20. Jahrhundert – ein Überblick', in B. Lemke (ed.), *Luft- und Zivilschutz in Deutschland im 20. Jahrhundert*, Potsdam: Militärgeschichtliches Forschungsamt, 67–88.

Lemke, T. (2011) *Biopolitics: An Advanced Introduction*. New York and London: NYU Press.

Lennartz, W. (1958) 'Luftschutzaufklärung und ihre Resonanz', *Ziviler Luftschutz*, 21(12), pp. 284–288.

Lenoir, T. (2000) 'All but war is simulation: The military entertainment complex', *Configurations*, 8, pp. 238–335.

Leshem, N. and Pinkerton, A. (2016) 'Re-inhabiting no-man's land: Genealogies, political life and critical agendas', *Transactions of the Institute of British Geographers*, 41, pp. 41–53.

Leutz, H. (1956) 'Aufgaben des zivilen Luftschutzes', *Ziviler Luftschutz*, 20(12), pp. 333–340.

Löfken, A. (1953) 'Grossraumplanung und Luftschutz', *Ziviler Luftschutz*, 18(12), pp. 282–283.

Löfken, A. (1954) 'Warum "Baulicher Luftschutz?"', *Ziviler Luftschutz*, 17, 2, 33–34.

Löfken, A. (1960) 'Überleben in Schutzraumbauten', *Ziviler Luftschutz*, 24, 5, 162–168.

Loriaux, M. (2008) *European Union and the Deconstruction of the Rhineland Frontier*. Cambridge: Cambridge University Press.

Luber, B. (1982) *Bedrohungsatlas Bundesrepublik Deutschland*. Wuppertal: Jugenddienst.

Luftwaffe (2011) 'Nachrichten des Jahres 2011', available at http://www.luftwaffe.de/portal/a/luftwaffe/!ut/p/c4/04_SB8K8xLLM9MSSzPy8xBz9CP3I5EyrpHK9nHK9_KJ0vZzStJJcvazEpPx0Y2O9xKLkjMwyPSMDQ0P9gmxHRQCb_bhf (accessed 30 October 2014).

Luftwaffe (2014) 'Nachrichten des Jahres 2011', available at http://www.luftwaffe.de/portal/a/luftwaffe/!ut/p/c4/04_SB8K8xLLM9MSSzPy8xBz9CP3I5EyrpHK

9nHK9_KJ0vZzStJJcvazEpPx0Y2O9xKLkjMwyPSMDQxP9gmxHRQAobHUP (accessed 30 October 2014).

Luke, T. and Ó Tuathail, G. (2000) 'Thinking geopolitical space: The spatiality of war, speed and vision in the work of Paul Virilio', in M. Crang (ed.), *Thinking Space*. London: Routledge, 360–397.

MacDonald, F. (2006a) 'The last outpost of empire: Rockall and the Cold War', *Journal of Historical Geography*, 32, pp. 627–647.

MacDonald, F. (2006b) 'Geopolitics and 'the Vision Thing': Regarding Britain and America's first nuclear missile', *Transactions of the Institute of British Geographers*, 31, pp. 53–71.

MacDonald, F. (2008) 'Space and the atom: On the popular geopolitics of Cold War rocketry', *Geopolitics*, 13(4), pp. 611–634.

MacDonald, F., Hughes, R. and Dodds, K. (2010) *Observant States: Geopolitics and Visual Culture*. London: I. B. Tauris.

Mackinder, H. (1904) 'The geographical pivot of history', *The Geographical Journal*, 13(4), pp. 421–444. (Reprinted 2004, *The Geographical Journal*, 170(4), pp. 298–321.

Mackinder, H. (1908) *The Rhine: Its Valley and Its History*. London: Chatto and Windus.

Mamadouh, V. (2005) 'Geographers and war, geographers and peace', in C. Flint (ed.), *The Geography of War and Peace: From Death Camps to Diplomats*. Oxford: Oxford University Press, pp. 26–60.

Mamadouh, V. (2014) 'One union, two speakers, three presidents, and ... 500 million EU citizens: The European Union and the 2012 Nobel Peace Prize', *Political Geography*, 42, A1–A3.

Marcuse, H. (1964[1999]) *One-Dimensional Man: Studies in the Ideology of Advanced Industrial Society*. London: Routledge.

Markusen, E. and Kopf, D. (1995) *The Holocaust and Strategic Bombing: Genocide and Total War in the 20th Century*. Boulder: Westview.

Masco, J. (2006) *The Nuclear Borderlands: The Manhattan Project in Post-Cold War New Mexico*. Princeton: Princeton University Press.

Masco, J. (2008) 'Survival is your business: Engineering ruins and affect in nuclear America', *Cultural Anthropology*, 23(2), pp. 361–398.

Masco, J. (2009) 'Life underground: Building the bunker society', *Anthropology Now*, 1(2), pp. 13–29.

Masco, J. (2014) *The Theatre of Operations: National Security Affect from the Cold War to the War on Terror*. Durham and London: Duke University Press.

Massey, D. (2005) *For Space*. London: Sage.

Mastny, V. and Byrne, M. (2006) *A Cardboard Castle? An Inside History of the Warsaw Pact*. Budapest and New York: CEU Press.

Matless, D., Oldfield, J. and Swain, A. (2008) 'Geographically touring the eastern bloc: British Geography, travel cultures and the Cold War', *Transactions of the Institute of British Geographers*, 33(3), pp. 354–375.

Maull, O. (1951) 'Europa – nicht Erdteil sondern Aufgabe', *Zeitschrift für Geopolitik*, 22, 11, 666–670.

Maull, O. (1956) *Politische Geographie*. Berlin: Safari.

Mazower, M. (2008) 'Foucault, Agamben: Theory and the Nazis', *Boundary 2*, 35(1), 23–34.

Mbembe, A. (2003) 'Necropolitics', *Public Culture*, 15(1), pp. 11–40.

McCormack, D. (2012) 'Geography and abstraction: Towards an affirmative critique', *Progress in Human Geography*, 36, pp. 715–734.

Megoran, N. (2006) 'For ethnography in political geography: Experiencing and re-imagining Ferghana Valley boundary closures', *Political Geography*, 25, pp. 622–640.

Megoran, N. (2011) 'War and peace? An agenda for peace research and practice in geography', *Political Geography*, 30, pp. 178–189.

Metz, D. (1955) 'Atomwaffenmanöver der Natotruppen 1954 in Westdeutschland', *Ziviler Luftschutz*, 19(12), pp. 62–64.

Michaud, E. (1993) 'National Socialist architecture as an acceleration of time', *Critical Inquiry* 19(2), pp. 220–233.

Michel, B. (2016) '"With almost clean or at most slightly dirty hands". On the selfdenazification of German geography after 1945 and its rebranding as a science of peace', *Political Geography*, 55, pp. 135–143.

Mielenz, W. (1953) 'Technisch-wissenschaftliche Probleme des zivilen Luftschutzes', *Ziviler Luftschutz*, 17(1), pp. 5–6.

Miksche, O. (1955) *Atomwaffen und Streitkräfte*. Bonn: Verlag Westunion.

Minca, C. (2005) 'The return of the camp', *Progress in Human Geography*, 29(4), pp. 405–412.

Minca, C. (2006) 'Giorigio Agamben and the new biopolitical *nomos*', *Geografiska Annaler* 88B(4), pp. 387–403.

Minca, C. (2015) 'Geographies of the camp', *Political Geography*, 49, pp. 74–83.

Minca, C. and Giaccaria, P (2016) *Hitler's Geographies*, Chicago: University of Chicago Press.

Minca, C. and Rowan, R. (2014) 'Guest editorial: The trouble with Schmitt', *Political Geography*, 41, A1–A3.

Minca, C. and Rowan, R. (2015a) *On Schmitt and Space*. London: Routledge.

Minca, C. and Rowan, R. (2015b) 'The question of space in Carl Schmitt', *Progress in Human Geography*, 39(3), pp. 268–289.

Ministerium für Nationale Verteidigung (1959) 'Bericht über Depots in Westdeutschland, die sich für eine Lagerung von Kernmunition eignen', 1 June 1962, Militärarchiv Freiburg DVW 1/25870/e.

Ministerium für Nationale Verteidigung (1964) 'Kurzfassung "Bericht über die Kampfbereitschaft, den Ausbildungsstand, und die Dislozierung der Kernwaffeneinheiten der NATO-Land- und Luftstreitkräfte in Westdeutschland', 15 June 1964, Militärarchiv Freiburg DVW 1/25784/e.

Ministerium für Nationale Verteidigung (1972) 'Informationen über die Aufklärungsmerkmale für ortsfeste Kernmittellage der NATO-Streitkräfte', 20 August 1972, Militärarchiv Freiburg DVW 1/42631.

Ministerium für Staatssicherheit (1966a) 'Sonderbericht Nr. 2/66 der Verwaltung Aufklärung des MfNV', 1 November 1966, *The Federal Commissioner for the Stasi Archives*, MfS ZAIG Nr. 6630.

Ministerium für Staatssicherheit (1966b) 'Einzel-informationen über Erfahrungsbericht des Bundesverteidigungsministeriums über die NATO Übung "Fallex 66"', 1 July 1966, The *Federal Commissioner for the Stasi Archives*, MfS Hauptverwaltung Aufklärung Nr. 225.

Ministerium für Staatssicherheit (1966c) 'Fallex 66 – oder Aggressionskrieg heute', no date, *The Federal Commissioner for the Stasi Archives*, MfS HA IX/MF/11992.

Ministerium für Staatssicherheit (1966d) 'Notstandsregelungen bei "Fallex 66"', 7 November 1966, *The Federal Commissioner for the Stasi Archives*, MfS ZAIG/1 9040.

Ministerium für Staatssicherheit (1967) 'Dokumentation über die Ausweichführungspunkte der Bonner Regierung, des Bundesinnenministeriums und des Bundesverteidigungsministeriums in der Eifel', 23 February 1967, *The Federal Commissioner for the Stasi Archives*, MfS 060 Nr.75/67.

Molitor, J. (2011) *Mit der Bombe überleben: Die Zivilschutzliteratur in der Bundesrepublik 1960–1964*. Marburg: Tectum.

Molitor, J. (2015) 'Die totale Verteidigung? Zivilschutz – aus zeithistorischer Perspektive', *Zeitschrift für Außen- und Sicherheitspolitik*, 8, pp. 389–405.

Monteyne, D. (2011) *Fallout Shelter: Designing for Civil Defence in the Cold War*. Minneapolis: University of Minnesota Press.

Morgenthau, H. (1948[1993]) *Politics Among Nations: The Struggle for Power and Peace*. New York: Knopf.

Müller, G. H. (1996) *Friedrich Ratzel (1844–1904): Naturwissenschaftler, Geograph, Gelehrter*. Stuttgart: GNT Verlag.

Müller, M. (2008) 'Reconsidering the concept of discourse for the field of critical geopolitics: Towards discourse as language and practice', *Political Geography*, 27, pp. 322–338.

Müller, M. (2013) 'Lack and jouissance in hegemonic discourse of identification with the state', *Organization*, 20, 279–298.

Murphy, A. and Johnson, C. (2004) 'German geopolitics in transition', *Eurasian Geography and Economics* 45(1), pp. 1–17.

Murphy, D. T. (1997) *Heroic Earth: Geopolitical Thought in Weimar Germany 1918–1933*. Kent and London: The Kent State University Press.

Murphy, D.T. (2014) 'Hitler's geostrategist?: The myth of Karl Haushofer and the "Institut für Geopolitik"', *The Historian*, 76(1), pp. 1–25.

Nast, H. (2003) 'Oedipalizing geopolitics: A commentary on Condensing the Cold War', *Geopolitics*, 8, pp. 190–196.

Nedelmann, C. (1985) 'Von der deutschen Minderwertigkeit', in C. Nedelmann (ed.), *Zur Psychoanalyse der nuklearen Drohung*. Göttingen: Verlag für medizinische Psychologie, pp. 11–34.

Nehring, H. (2013) *Politics of Security: The British and West German Protests Against Nuclear Weapons and the Early Cold War, 1945–1970*. Oxford: Oxford University Press.

Netz, R. (2004) *Barbed Wire: A Political Ecology of Modernity*. Middletown: Wesleyan University Press.

Neues Deutschland (1966a) '… und Lücke befahl den Notstand', 23 October 1966.

Neues Deutschland (1966b) 'Fallex 66 Westdeutsche Städte im Kernwaffenhagel', no date, The *Federal Commissioner for the Stasi Archives*, MfS ZAIG 9040.

No author (1955) 'Wasserstoffbombe und göttliche Vorsehung… "um den Begriff zu gebrauchen"' *Zeitschrift für Geopolitik*, 26(1), p. 1.

No author (1956) 'An unsere Leser!', *Zeitschrift für Geopolitik in Gemeinschaft und Politik*, 27(3), pp. 55–56.

No author (1959) 'Kurzbericht des Bundesamtes für zivilen Bevölkerungsschutz über die im Auftrage des BMWo durchgeführten Belegungsversuche eines luftstoßsicheren Schutzbaues in Waldbröl vom 9.-10. 1. 1959 und vom 26.-31.1.1959', *Ziviler Luftschutz*, 23(6), pp. 167–169.

No author (1962) *Atomgefahren: Was stimmt? Was kommt? Was tun?* Seebruck: Heering.

No author (1966) 'Tatort Ahrweiler', *The Federal Commissioner for the Stasi Archives*, MfS ZAIG, p. 11114.

Northern Army Group (1966) 'COMNORTHAGS post-exercise report', 7 December 1966, Military Archives Freiburg 7–1/27.

Ó Tuathail, G. (1994) '(Dis)placing geopolitics: Writing on the maps of global politics', *Environment and Planning D*, 12, pp. 525–546.

Ó Tuathail, G. (1996) *Critical Geopolitics: The Politics of Writing Global Space*. Minneapolis: University of Minnesota Press.

Ó Tuathail, G. and Agnew, J. (1992) 'Geopolitics and discourse: Practical geopolitical reasoning in American foreign policy', *Political Geography*, 11(2), pp. 190–204.

Ó Tuathail, G. and Dalby, S. (1998) 'Introduction: Rethinking geopolitics: Towards a critical geopolitics', in G. Ó Tuathail and S. Dalby (eds), *Rethinking Geopolitics*. London: Routledge, pp. 1–15.

Paetsch, H. (1952) 'Luftschutz einst und jetzt', *Ziviler Luftschutz*, 16, 1, 6–10

Pain, R. (2016) 'Intimate war', *Political Geography*, 44, pp. 64–73.

Pain, R. and Smith, S. (2008) *Fear: Critical Geopolitics and Everyday Life*. Aldershot: Ashgate.

Panorama (2017a) 'Donald Trump und US-Atombomben in Deutschland', available at https://daserste.ndr.de/panorama/archiv/2017/US-Atombomben-in-Deutschland-und-Donald-Trump,atombombe100.html (accessed 12 February 2017)

Panorama (2017b) 'Stellungnahme: Donald Trump und US-Atombomben in Deutschland', available at https://daserste.ndr.de/panorama/aktuell/Stellungnahme-Donald-Trump-und-US-Atombomben-in-Deutschland,atombombe104.html (accessed 12 February 2017).

Pantenius, H. J. (1952) 'Die Bedeutung der Flugzeugträgers', *Zeitschrift für Geopolitik*, 23(10), pp. 582–592.

Pfeffer, K. H. and Vowinckel, K. (1951) 'In eigener Sache', *Zeitschrift für Geopolitik*, 22(1), pp. 80.

Pfeffer, K. H. and Vowinckel, K. (1954) 'Was will Geopolitik heute? *Zeitschrift für Geopolitik*, 24(4), pp. 193–200.

Pile, S. (1996) *The Body and the City: Psychoanalysis, Space and Subjectivity*. London and New York: Routledge.

Pile, S. (2001) 'The unknown city … or, an urban geography of what lies buried below the surface', in I. Borden et al. (eds), *The Unknown City: Contesting Architecture and Urban Space*. Cambridge: MIT Press, pp. 262–279.

Pohl, M. (1999) *Philipp Holzmann: Geschichte eines Bauunternehmens 1849–1999*. München: C. H. Beck.

Power, M. (2007) 'Digitized virtuosity: Video war games and post-9/11 cyberdeterrence', *Security Dialogue*, 38, pp. 271–88.

Preute, M. (1984) *Vom Bunker der Bundesregierung*. Köln: Edition Nachtraben.

Ramadan, A. (2009) 'Destroying Nahr el-Bared: Sovereignty and urbicide in the space of exception', *Political Geography*, 28, pp. 153–163.

Ramadan, A. (2013) 'Spatialising the refugee camp', *Transactions of the Institute of British Geographers*, 38, pp. 65–77.

Ramcke, H. B. (1951) 'Dem Marschall von Frankreich', *Zeitschrift für Geopolitik*, 22(8), pp. 469–470.

Ramcke, H. B. (1952) 'Wie wird man Kriegsverbrecher?', *Zeitschrift für Geopolitik*, 23(11), pp. 640–641.

Raskin, M. (1982) 'Nuclear exterination and the national security state', in *New Left Review* (eds), *Exterminism and Cold War*. London: Verso, pp. 205–222.

Ratzel, F. (1876[1988]) *Sketches of Urban and Cultural Life in North America*. New Brunswick and London: Rutgers University Press.

Ratzel, F. (1882[1909]) *Anthropogeographie Erster Teil: Grundzüge der Anwendung der Erdkunde auf die Geschichte*. Stuttgart: Engelhorn.

Ratzel, F. (1897) *Politische Geographie*. München und Leipzig: R. Oldenbourg.

Ratzel, F. (1901) 'Der Lebensraum: eine biogeographische Studie', in K. Bücher et al. (eds), *Festgaben für Albert Schäffle zur siebenzigen Wiederkehr seines Geburtstages*. Tübingen: Lapp, 102–189.

Ratzel, F. (1905[1966]) *Jugenderinnerungen*. München: Kösel.

Ratzel, F. (1906) *Kleinere Schriften von Friedrich Ratzel, herausgegeben von Hans Helmolt*. München: R Oldenbourg.

Ratzel, F. (1911) *Deutschland: eine Einführung in die Heimatkunde*. Berlin: Verlag von Georg Reimer.

Ratzel, F. (1941) *Erdenmacht und Völkerschicksal, herausgegeben und eingeleitet von Generalmajor a. D. Prof. Dr. Karl Haushofer*. Stuttgart: Alfred Kröner Verlag.

Reuber, P. (2009) *Politische Geographie*. Paderborn: Ferdinand Schöningh.

Reuber, P. and Wolkersdorfer, G. (2002) 'The transformation of Europe and the German contribution – critical geopolitics and geopolitical representations', *Geopolitics*, 7(3), pp. 39–60.

Reid, J. (2008) 'Life struggles: War, discipline and biopolitics in the thought of Michel Foucault', in M. Dillon and A. W. Neal (eds), *Foucault on Politics, Security and War*. Basingstoke: Palgrave, pp. 65–92.

Rose, K. D. (2001) *One Nation Underground*. New York: New York University Press.

Ross, A. (2011) 'Geographies of war and the putative peace', *Political Geography*, 30, pp. 197–199.

Ruge, F. (1955a) 'Die vergessene See', *Zeitschrift für Geopolitik*, 26(6), pp. 355–360.

Ruge, F. (1955b) *Seemacht und Sicherheit: Eine Schicksalsfrage für alle Deutschen*. Tübingen: Verlag Fritz Schlichtenmeyer.

Ruge, F. (1963) *Politik, Militär, Bündnis*. Stuttgart: Deutsche Verlagsanstalt.

Rutherford, P. and Rutherford, S. (2013) 'The confusions and exuberances of biopolitics', *Geography Compass*, 7(6), pp. 412–422.

Said, E. (1995) *The Politics of Dispossession: The Struggle for Palestinian Self-Determination, 1969–1994*. New York: Pantheon.

Salter, M. B. (2011a) 'Gaming world politics: Meaning of play and world structure', *International Political Sociology*, 54, pp. 453–456.

Salter, M. B. (2011b) 'The geographical imaginations of video games diplomacy, civilization, America's Army and Grand Theft Auto IV', *Geopolitics*, 16, pp. 359–388.

Samhaber, E. (1952) 'Die Verteidigung Westdeutschlands', *Zeitschrift für Geopolitik*, 23(11), p. 649–654.

Sandner, G. (2000) 'Wiederbegegnung nach 40 Jahren: Peter Schöller und der Start der Auseinandersetzung der Geographie mit der Geopolitik im "Dritten Reich"', in I. Diekmann, Peter Krüger and J. H. Schoepps (eds), *Geopolitik – Grenzgänge im Zeitgeist*. Potsdam: Verlag Berlin Brandenburg, pp. 403–418.

Scharpff, A. (1955) 'Lebensräume und Ländergrenzen', *Zeitschrift für Geopolitik*, 26(7), pp. 392–394.

Schlögel, K. (2011) (revised edition) *Im Raume lesen wir die Zeit; Über Zivilisationsgeschichte und Geopolitik*. Frankfurt: Fischer.

Schmidle, A. (1959) 'Schutzraumbau für die Bevölkerung muß "das Primäre" aller Luftschutzmaßnahmen sein', *Ziviler Luftschutz*, 23(9), pp. 253–255.

Schmitt, C. (1922[2005]) *Political Theology: Four Chapters On the Concept of Sovereignty*. Chicago and London: University of Chicago Press.

Schmitt, C. (1932[2007]) *The Concept of the Political*. Chicago and London: University of Chicago Press.

Schmitt, C. (1941[1991]) *Völkerrechtliche Großraumordnung mit Interventionsverbot für raumfremde Mächte*. Berlin: Duncker and Humblot.

Schmitt, C. (1942[2001]) *Land und Meer*. Stuttgart: Klett-Cotta.

Schmitt, C. (1950[1997]) *Der Nomos der Erde im Völkerrecht des Jus Publikum Europaeum*. Stuttgart: Klett Cotta.

Schmitt, C. (1952) 'Zum Gedächtnis von Sege Maiwald', *Zeitschrift für Geopolitik*, 23(7), pp. 447–448.

Schmitthenner, H. (1951) *Lebensräume im Kampf der Kulturen*. Heidelberg: Quelle and Meyer.

Schnitzer, E. W. (1955) 'German geopolitics revived', *The Journal of Politics*, 17, pp. 407–423.

Schöller, P. (1957) 'Wege und Irrwege der Politischen Geographie und Geopolitik', *Erdkunde*, 11, pp. 1–20.

Schregel, S. (2011) *Der Atomkrieg vor der Wohnungstür: eine Politikgeschichte der neuen Friedensbewegung in der Bundesrepublik 1970–1985*. Frankfurt/New York: Campus.

Schröder, G. (1954) 'Hat Luftschutz noch einen Sinn?', *Ziviler Luftschutz*, 18(5), pp. 109–110.

Schulze-Hinrichs, A. (1957) 'Die militärgeographische Lage', in J. Rohwer (ed.), *Seemacht heute*. Oldenburg and Hamburg: Stalling, pp. 33–43.

Schützsack, U. (1955) 'Die Stellung der Frau in der Zivilverteidigung', *Ziviler Luftschutz*, 19(5), pp. 121–123.

Schützsack, U. (1960) 'Die Ruinenstadt in Tinglev: Zentrale Ausbildungsstätte des dänischen Rettungsdienstes', *Ziviler Luftschutz*, 24(3), pp. 79–82.

Searle, A. (1998) 'A very special relationship: Basil Liddell Hart, Wehrmacht Generals and the debate on West German rearmament, 1945–1953', *War in History*, 5(3), pp. 327–357.

Searle, A. (2003) *Wehrmacht Generals, West German Society, and the Debate on Rearmament, 1949–1959*. Westport: Praeger.

Sebald, W. G. (1999) *On the Natural History of Destruction*. London Hamish: Hamilton.

Segal, H. (1995) 'From Hiroshima to the Gulf War and after: A psychoanalytic perspective', in A. Elliot and S. Frosh (eds), *Psychoanalysis in Contexts: Paths Between Theory and Modern Culture*. London, Routledge, pp. 191–204.

Segal, H. (1997) *Psychoanalysis, Literature and War: Papers 1972–1995*. London: Routledge.

Semple, E. C. (1911) *Influences of Geographic Environment on the Basis of Ratzel's System of Anthropo-Geography*. New York: Holt.

SHAPE (1974) 'Storage and utilization of nuclear weapons in Allied command Europe (ACE) (NU)', 1 November 1974, Militärarchiv Freiburg BW 1/103524.

Sharp, J. (1993) 'Publishing American identity: Popular geopolitics, myth and the Reader's Digest', *Political Geography*, 12(6), pp. 491–503.

Sharp, J. (2000a) 'Re-masculinising geo-politics? Comments on Gearoid Ó Tuathail's Critical Geopolitics', *Political Geography*, 19, 361–364.

Sharp, J. (2000b) *Condensing the Cold War: Reader's Digest and American Identity*. Minneapolis: University of Minnesota Press.

Shaw, I. G. R. (2010) 'Playing war', *Social and Cultural Geography*, 11, pp. 789–803.

Shaw, I. G. R. and Sharp, J. (2013) 'Playing with the future: Social irrealism and the politics of aesthetics', *Social and Cultural Geography*, 14, pp. 341–359.

Shaw, M. (2008) 'New wars of the city: Relationships of "urbicide" and "genocide"', in S. Graham (ed.), *Cities, War, and Terrorism: Towards an Urban Geopolitics*. Malden: Blackwell Publishing, pp. 141–153.

Sloterdijk, P. (2009) *Terror from the Air*. Los Angeles: Semiotext(e).

Smith, N. (2003) *The Endgame of Globalisation*. London: Routledge.

Smith, N. (2004) *American Empire: Roosevelt's Geographer and the Prelude to Globalization*, Oakland: University of California Press.

Smith, W. (1980) 'Friedrich Ratzel and the origins of Lebensraum', *German Studies Review*, 3(1), pp. 51–68.

Social and Cultural Geography (2003) 'Special Issue on psychoanalytic geographies', 43, pp. 283-399.

Soxhlet, E. (1953) 'LS-Bunker und Atombombe', *Ziviler Luftschutz*, 17(10), pp. 233–234.

Sparke, M. (2006) 'A neoliberal nexus: economy, security and the biopolitics of citizenship on the border', *Political Geography*, 25(2), pp. 151–180.

Speer, A. (1970) *Inside the Third Reich*. London: Phoenix.

Speidel, H. (1960) 'Die Bedeutung der Landstreitkräfte in der NATO', in G. Blumentritt (ed.), *Strategie und Taktik*. Konstanz: Akademische Verlagsgesellschaft Athenaion, pp. 159–169.

Speidel, H. (1969) *Zeitbetrachtungen: Ausgewählte Reden*. Mainz: Hase und Koehler.

Sprengel, R. (1996) *Kritik der Geopolitik: Ein deutscher Diskurs*. Berlin: Akademie.

Squire, V. (2015) 'Reshaping critical geopolitics? The materialist challenge', *Review of International Studies*, 41, pp. 139–159.

Stadtverwaltung Ahrweiler (1967) 'Wasserrechtliche Genehmigung zur Einleitung geklärter Abwässer in den Kratzenbach', Federal Archives Koblenz B157/3841.

Stahl, R. (2010) *Militainment, Inc.: War, Media, and Popular Culture*. London, Routledge.

Stead, N. (2003) 'The value of ruins: Allegories of destruction in Benjamin and Speer', *Form/Work: An Interdisciplinary Journal of the Built Environment*, 6, pp. 51–64.

Steinberg, P. and Peters, K. (2015) 'Wet ontologies, fluid spaces: Giving depth to volume through oceanic thinking', *Environment and Planning D*, 33(2), pp. 247–264.

Stenck, N. (2008) 'Eine verschüttete Nation? Zivilschutzbunker in der Bundesrepublik Deutschland 1950–1965', in I. Marszolek and M. Buggeln (eds), *Bunker: Kriegsort, Zuflucht, Erinnerungsraum*. Frankfurt/New York: Campus, 75–87.

Strausz-Hupé, R. (1942) *Geopolitics: The Struggle for Space and Power*. New York: J. P. Putnam's Sons.

Süddeutsche Zeitung (1985) 'Im Bunker den Papier-Krieg überstehen', 20 February 1985.

Svirsky, M. and Bignall, S. (2012) *Agamben and Colonialism*. Edinburgh: University of Edinburgh Press.

Tagesspiegel (2017) 'Deutschland braucht Atomwaffen', available at https://causa. tagesspiegel.de/politik/europa-und-die-weltweiten-krisen/deutschland-braucht-atomwaffen.html (accessed 12 January 2017).

Taylor, R. (1974) *Word in Stone: The Role of Architecture in the National-Socialist Ideology*. Berkeley: University of California Press.

Thies, J. (2012) *Hitler's Plans for Global Domination: Nazi Architecture and Ultimate War Aims*. New York and Oxford: Berghahn.

Thompson, E. P. (1982a) 'Notes on exterminism: The last stage of civilisation', in *New Left Review* (eds), *Exterminism and Cold War*. London: Verso, pp. 1–33.

Thompson, E. P. (1982b) *Zero Option*. London: Merlin.

Thoß, B. (2007) 'Der Regierungsbunker in der Strategie der atomaren Abschreckung Bundesamt für Bauwesen und Raumordnung', (ed.) *Der Regierungsbunker* (Berlin und Tübingen Ernst Wasmuth Verlag), pp. 32–41.

Thrift, N. (2000) 'It's the little things', in K. Dodds and D. Atkinson (eds), *Geopolitical Traditions: A Century of Geopolitical Thought*. London: Routledge, pp. 380–387.

Thüringer Allgemeine (2015) 'Mythos Bunker Rothenstein: Fragen über Fragen zum einst größten Untertage-Waffendepot in Zentraleuropa', available at http://www. thueringer-allgemeine.de/web/zgt/leben/detail/-/specific/Mythos-Bunker-Rothenstein-Fragen-ueber-Fragen-zum-einst-groessten-Untertage-Waf-171024120 (accessed 15 August 2016).

Till, K. (2005) *The New Berlin: Memory, Politics, Place*. Minneapolis: University of Minnesota Press.

Toal, G. (2017) *Near Abroad: Putin, the West and the Contest over Ukraine and the Caucasus*. Oxford: Oxford University Press.

Tribüne (1967) 'Süßes Leben im Notstandsbunker', 10 March 1967.

Troll, C. (1949) 'Geographic science in Germany during the period 1933–1945: A critique and justification', *Annals of the Association of American Geographers*, 39(2), pp. 99–137.

Tunander, O. (2001) 'Swedish-German geopolitics for a new century: Rudolf Kjellén's "The State as a Living Organism"', *Review of International Studies*, 27(3), pp. 451–463.

UK Ministry of Defence (1967) 'Fallex 66 – Post exercise report', 20 January 1967, The National Archives, Kew CAB 164/77.

US Army (1984) 'Manual: Special ammunition (nuclear) direct and general support units operations', 28 December 1984, Militärarchiv Freiburg BWD 5/1702.

US Army Headquarters Europe (1964) 'Letter: Change to multinational technical arrangement', 13 March 1964, Militärarchiv Freiburg BW/1 166944/b.

USCINCEUR (1975) 'Fernschreiben: Consideration of hostages', 15 December 1975, Militärarchiv Freiburg BW 1/106557.

Vanderbilt, T. (2002) *Survival City: Adventures Among the Ruins of Atomic America*. New York: Princeton Architectural Press.

Vaughan-Williams, N. (2015) *Europe's Border Crisis: Biopolitical Security and Beyond*. Oxford: Oxford University Press.

Virilio, P. (1975) *Bunker Archaeology*. New York: Princeton Architectural Press.

Virilio, P. (2012) *The Administration of Fear*. Los Angeles: Semiotext(e).

Virilio, P. and Lothringer, S. (1983) *Pure War*. New York: Semiotext(e).

Volksblatt Berlin (1985) 'NATO "spielt" dritter Weltkrieg', 1 March 1985.

von Hilgers, P. (2012) *War Games: A History of War on Paper*. Cambridge and London: MIT Press.

Von Schweppenburg, L. (1952a) *Gebrochenes Schwert*. Homburg: Bernhard und Graefe.

Von Schweppenburg, L. (1952b) *Die große Frage: Gedanken über die Sowjetmacht*. Homburg: Bernhard und Graefe.

Vulpius, A. (1964) 'Bemerkungen zur Öffentlichkeitsarbeit', *Ziviler Bevölkerungsschutz*, 9(2), pp. 16–17.

Waltz, K. (1979) *Theory of International Politics*. New York: McGraw-Hill.

Wanklyn, H. (1961) *Friedrich Ratzel: A Biographical Memoir and Bibliography*. Cambridge: Cambridge University Press.

Wassermann, F. M. (1952) 'Karl Haushofer', *Zeitschrift für Geopolitik*, 23(12), pp. 721–726.

Weinstein, A. (1951) *Armee ohne Pathos: Die deutsche Wiederbewaffnung im Urteil ehemaliger Soldaten*. Bonn: Köllen Verlag.

Weizsäcker, C. v. (1952) 'Albrecht Haushofer', *Zeitschrift für Geopolitik*, 23(4), pp. 193–195.

Weizman, E. (2002) 'Introduction to the politics of verticality', 23 April 2002, available at https://www.opendemocracy.net/ecology-politicsverticality/article_801.jsp (accessed 6 December 2015).

Weizman, E. (2007) *Hollow Land: Israel's Architecture of Occupation*. London: Verso.

Welt im Bild (1953) 'Welt im Bild: Die Allianz Wochenschau 70/53', 28 October 1953, available at https://www.filmothek.bundesarchiv.de/video/583133 (accessed 21 July 2015).

Werber, N. (2005) 'Geo- and biopolitics of Middle-earth: A German reading of Tolkien's The Lord of the Rings', *New Literary History*, 36(2), pp. 227–246.

Werber, N. (2014) *Geopolitik zur Einführung*. Hamburg: Junus.

Westfälische Rundschau (1961) 'Letzte Rettung: Der Betonklotz', 15 December 1961.

Williams, M. C. (2005) *The Realist Tradition and the Limits of International Relations*. Cambridge: Cambridge University Press.

Williams, P. and McConnell, F. (2011) 'Critical geographies of peace', *Antipode*, 43, pp. 927–931.

Wills, J. (2001) '"Welcome to the Atomic Park": American nuclear landscapes and the "unnaturally natural"', *Environment and History*, 7, pp. 449–472.

Wills, J. (forthcoming) 'Exploding the 1950s consumer dream: Mannequins and mushroom clouds at Doom Town, Nevada Test Site' (under review).

Wilson, R. M. (2015) 'Mobile bodies: Animal migration in North American history', *Geoforum*, 65, pp. 465–472.

Wirtschaftswoche (2017) 'Über Atomwaffen reden', available at http://www.wiwo.de/my/politik/ausland/nukleare-kooperation-ueber-atomwaffen-reden-/19366768.html (accessed 12 January 2017).

Woodward, R. (2004) *Military Geographies*. Malden: Blackwell.

Woodward, R. (2014) 'Military landscapes: Agendas and approaches for future research', *Progress in Human Geography*, 38(1), pp. 40–61.

Woodyer, T. (2012) 'Ludic geographies: Not merely child's play', *Geography Compass*, 66, pp. 313–326.

Woodyer, T. and Geoghegan, H. (2012) '(Re)enchanting geography? The nature of being critical and the character of critique in human geography', *Progress in Human Geography*, 37, pp. 195–214.

Yarwood, R. (2015) 'Miniaturisation and the representation of military geographies in recreational wargaming', *Social and Cultural Geography*, 16(6), pp. 654–674.

ZDv 49/20 (1961) 'Sanitätsausbildung aller Truppen (Lehrschrift)', December 1961, Federal Archives Koblenz B 106/85615.

Zehfuss, M. (2007) *Wounds of Memory: The Politics of War in Germany*. Cambridge: Cambridge University Press.

Zimmermann, P. (1958) 'Grundsätzliche Gedankengänge über die Entwicklung der Schutzraumausstattung', *Ziviler Luftschutz*, 21(6), pp. 135–139.

Žižek, S. (1991) *Looking Awry: An Introduction to Jacques Lacan Through Popular Culture*. Cambridge: MIT Press.

# Index

*Cryptic Concrete: A Subterranean Journey Into Cold War Germany*, First Edition. Ian Klinke.
© 2018 John Wiley & Sons Ltd. Published 2018 by John Wiley & Sons Ltd.